Structure of the Earth

Structure of the Earth

Sydney P. Clark, Jr.

Yale University

Prentice-Hall, Inc., Englewood Cliffs, New Jersey

Illustrations by Richard Kassouf

PRENTICE-HALL INTERNATIONAL, INC., *London*
PRENTICE-HALL OF AUSTRALIA, PTY. LTD., *Sydney*
PRENTICE-HALL OF CANADA, LTD., *Toronto*
PRENTICE-HALL OF INDIA PRIVATE LIMITED, *New Delhi*
PRENTICE-HALL OF JAPAN, INC., *Tokyo*

Current printing (last digit):
10 9 8 7 6 5 4 3 2 1

FOUNDATIONS OF EARTH SCIENCE SERIES

A. Lee McAlester, Editor

(p) 13-854646-0
(c) 13-854653-3

To Dibbie, Teddy, Jordy, Lisa, Christie
and Kailua

Foundations

of Earth Science Series

Elementary Earth Science textbooks have too long reflected mere traditions in teaching rather than the triumphs and uncertainties of present-day science. In geology, the time-honored textbook emphasis on geomorphic processes and descriptive stratigraphy, a pattern begun by James Dwight Dana over a century ago, is increasingly anachronistic in an age of shifting research frontiers and disappearing boundaries between long-established disciplines. At the same time, the extraordinary expansions in exploration of the oceans, atmosphere, and interplanetary space within the past decade have made obsolete the unnatural separation of the "solid Earth" science of geology from the "fluid Earth" sciences of oceanography, meteorology, and planetary astronomy, and have emphasized the need for authoritative introductory textbooks in these vigorous subjects.

Stemming from the conviction that beginning students deserve to share in the excitement of modern research, the *Foundations of Earth Science Series* has been planned to provide brief, readable, up-to-date introductions to all aspects of modern Earth science. Each volume has been written by an authority on the subject covered, thus insuring a first-hand treatment seldom found in introductory textbooks. Four of the volumes—*Structure of the Earth, Earth Materials, The Surface of the Earth,* and *Earth Resources*—cover topics traditionally taught in physical geology courses. Four more volumes—*Geologic Time, Ancient Environments, The History of the Earth's Crust,* and *The History of Life*—treat historical topics. The remaining volumes—*Oceans, Man and the Ocean, Atmospheres, Weather,* and *The Solar System*—deal with the "fluid Earth" sciences of oceanography and atmospheric and planetary sciences. Each volume, however, is complete in itself and can be combined with other volumes in any sequence, thus allowing the teacher great flexibility in course arrangement. In addition, these compact and inexpensive volumes can be used individually to supplement and enrich other introductory textbooks.

Contents

1

Introduction   1

*Internal Divisions of the Earth.   The Surface of the Earth.
Volcanoes and Earthquakes.*

2

Geologic structures   9

*Faults and Folds. Nappes. Transcurrent Faults. Structures of
Some Other Mountain Systems.   Rift Valleys.*

3

The Earth's magnetic field   26

*The Dynamo Theory.   Rock Magnetism.   Paleomagnetism.   Other
Evidence for Continental Drift.   The Vine-Matthews
Hypothesis.*

4

Plate tectonics   46

Transform Faults.   A Closer Look at Ocean Ridges.   Ocean Trenches. Mountain Building.   Motions of Present-Day Plates.   The Driving Mechanism.

5

Seismology   67

Seismic Waves.   Free Oscillations of the Earth.   Travel Times of Body Waves.   Behavior of Body Waves at an Interface.   Travel Times and the Velocity Distribution in the Earth.   Seismic Attenuation.

6

**The constitution of the Earth
from seismic evidence   92**

Pressures in the Earth.   Increase of Density by Self-Compression of a Homogeneous Layer.   Inhomogeneity of the Earth's Mantle. Density Distribution in the Earth.   Constitution of the Crust and Upper Mantle.   Constitution of the Transition Zone and Lower Mantle.   Constitution of the Core.

7

The Earth's gravity field   105

Gravity Anomalies.   The Theory of Isostasy.   Departures from Isostatic Compensation.   The Shape of the Earth.

8

**Heat flow and the temperatures
in the Earth   118**

*Terrestrial Heat Flow.   Heat Generation by Radioactivity.
Temperatures Deep in the Earth.*

Suggestions for further reading   127

Index   128

1

Introduction

The Earth is a restless body. Most of the processes of change operate so slowly that they are imperceptible in a human lifetime. But they also operate inexorably, and given the billions of years embraced by geologic time they can produce profound effects. Great forces raise up mountains and distort the strata that were originally laid down as horizontal sheets. Other forces seem to move whole continents about, gradually rearranging the map of the world. Nevertheless, such processes are not always slow. Volcanic eruptions can build a mountain in a few years, and scarps associated with earthquakes are produced almost instantaneously.

The most widespread process that proceeds at a humanly perceptible rate is erosion. However, erosion is not the subject of this book; rather, it is covered in this series by A. L. Bloom in *The Surface of the Earth*. But the rapidity of erosion gives us some of our most direct evidence that internal forces are operating in the Earth today and have done so throughout geologic time. We can say this because at present there are high mountains on every continent. Erosion rates are such that mountains formed at the time of the creation of the Earth would have long since been worn away to inconspicuous hills. The existence of high mountains today and the strong evidence for their existence throughout the past show that forces responsible for the creation or rejuvenation of mountains have operated all during the Earth's history. Processes occurring in the Earth's interior are responsible for these forces, and the description of these processes and their effects forms the central theme of this book.

It is the immensity of geologic time that permits the slow changes to which the Earth is subject to have significant effects. Geologic time and the methods by which it is measured are covered in a companion volume of this series, *Geologic Time*, by D. L. Eicher. A summary of the geologic time scale is reproduced inside the back cover of the present book. The reader should consult Eicher's book for further details.

Much of our understanding of global dynamics stems from knowledge of the physical and chemical properties of rocks and minerals. The systematic classification and description of Earth materials is beyond the scope of this book; it too is covered in a companion volume, *Earth Materials*, by W. G. Ernst. Some of the terminology of mineralogy and petrology is used in the pages that follow. Should they be unfamiliar to the reader, he is referred to *Earth Materials* for further information.

Internal Divisions of the Earth

Scientists recognize three broad subdivisions of the Earth: the *crust*, a thin surface layer; the *mantle*, a much thicker layer beneath the crust; and the *core*, the central region of the Earth. The basis of these subdivisions is seismological data that is discussed at length in Chapter 5. The crust is separated from the mantle by a surface called, after its discoverer, the *Mohorovičić discontinuity*. This name is decidedly formidable, at least to English-speaking readers, and recent years have witnessed a merciful tendency towards abbreviation. We shall refer to it as the *Moho*. It is found at a greater depth beneath continents than beneath oceans. In the former regions it commonly lies about 40 km beneath the surface; in the latter, about 7 km beneath the ocean floor.

Some important properties of these subdivisions of the Earth are listed in Table 1–1, along with the same properties for the atmosphere and oceans. The mantle makes up about 80 per cent of the Earth's volume and has nearly 70 per cent of the total mass. The crust represents a thin veneer that is most insignificant when viewed on a global scale; the atmosphere and oceans are even less impressive.

We can use the densities given in Table 1–1 to form a preliminary no-

Table 1–1

Some Data on Major Subdivisions of the Earth

	Thickness or radius (km)	Mass (gm)	Mean density (gm/cm^3)
Oceanic crust	7	7.0×10^{24}	2.8
Continental crust	40	1.6×10^{25}	2.8
Mantle	2870	4.08×10^{27}	4.6
Core	3480	1.88×10^{27}	10.6
Oceans	4	1.39×10^{24}	1.0
Atmosphere	–	5.1×10^{21}	–
Earth	6371	5.98×10^{27}	5.5

tion of what the material making up the crust, the mantle, and the core might be. The mean density of the crust is well within the range exhibited by common rocks, and we therefore expect that the whole crust is similar chemically to what is exposed at the surface. There may be some change in composition with depth, but we can be fairly confident that it is minor. The density of the mantle is also consistent with rock material, but it is too high to be representative of common rocks. But there are comparatively rare rocks with higher densities than usual that are consistent with the mantle value. The core, on the other hand, is too dense to be rocky. It is believed to be a metal, probably mainly iron.

Despite their quantitative unimportance on a global scale, the crust, the oceans, and the atmosphere provide us with our home and the means to sustain life. More important, they represent nearly all of the material of the Earth that is available for direct study. Our knowledge of the much vaster interior of the Earth is restricted to indirect inferences. For this reason we start where our knowledge is most complete, at the Earth's surface.

The Surface of the Earth

The most obvious division of the Earth's surface is into areas of land and sea. This distinction is more than a purely superficial one; as we have already seen, the Earth's crust is thicker under continents than under oceans. Measurable differences in properties are now known to persist into the mantle to depths of a few hundred kilometers. The simple division into land and sea is rather a gross distinction and further detail can readily be perceived.

The locations of extremes of elevation on the Earth's surface are shown in the map of Fig. 1–1, which indicates the high plateaus and mountains

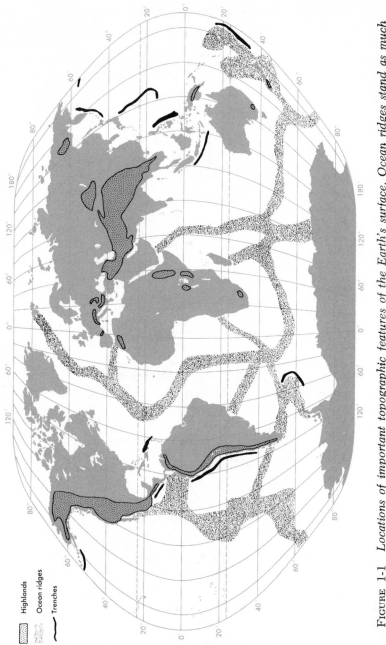

FIGURE 1-1 *Locations of important topographic features of the Earth's surface. Ocean ridges stand as much as 2 km above the surrounding abyssal plains, and water depths over their crests are commonly less than 4 km. Trenches are the deepest parts of the oceans, and water depths exceeding 10 km are attained.*

Highlands

Ocean ridges

Trenches

on land, and the great deeps and the shallow ridges in the sea. It is at once clear that the extremes of elevation are not located in the centers of continents and oceans but rather near their margins. The Cordilleran mountains of the Americas are a striking illustration of this fact. The Alpine-Himalayan chain is less clearly marginal to the continents, but it certainly is not centrally located either. The deepest parts of the oceans take the form of elongate trenches that are invariably close to land. They are concentrated in the northwest Pacific near arcuate chains of islands, which are termed *island arcs* because of their shape. Similarly, the great deeps in the Atlantic are adjacent to the island arcs of the Caribbean and Scotia Seas.

Elongate elevations of the sea floor, known as *rises* or *ridges*, are commonly found far from land. The best example is the Mid-Atlantic Ridge, which runs almost exactly up the center of the Atlantic Ocean. But, as Fig. 1–1 shows, these features occur in all the oceans. The East Pacific rise runs between Antarctica and Central America, and the Indian Ocean is split by the Mid-Indian and Carlsberg Ridges. These topographic highs on the sea floor are connected to each other and to the Mid-Atlantic Ridge by a series of rises in the Southern Ocean.

Volcanoes and Earthquakes

Indications of the regions in which geological activity is most pronounced at the present time are provided by volcanoes and earthquakes. Volcanoes occur where molten lava is generated at relatively shallow depths and is later poured out on the surface. Earthquakes are the result of the sudden release of energy that was stored up as a consequence of slow movement and deformation of the Earth's rocky material. Neither earthquakes nor volcanoes occur uniformly over the Earth's surface; rather, they are concentrated in certain regions. Both phenomena commonly occur together, indicating that regions where heat is available to melt rock are also regions where motions are most vigorous. We may speak of these as *active* zones.

The distribution of active volcanoes is shown in Fig. 1–2. The concentration of activity in the "girdle of fire" surrounding the Pacific Ocean is marked, but even within this girdle the distribution is not uniform. Volcanoes are clustered in the island arcs and portions of continents adjacent to the deep ocean trenches, as comparison with Fig. 1–1 shows. A weaker association between volcanoes and ocean ridges can also be perceived, particularly in the Atlantic and East Pacific Oceans. The activity of the ridges is probably underestimated because they are known to be the sites of submarine eruptions. These are recorded only if a ship happens to be in the vicinity when an eruption is taking place.

Other active volcanoes are scattered about in mountainous areas of East

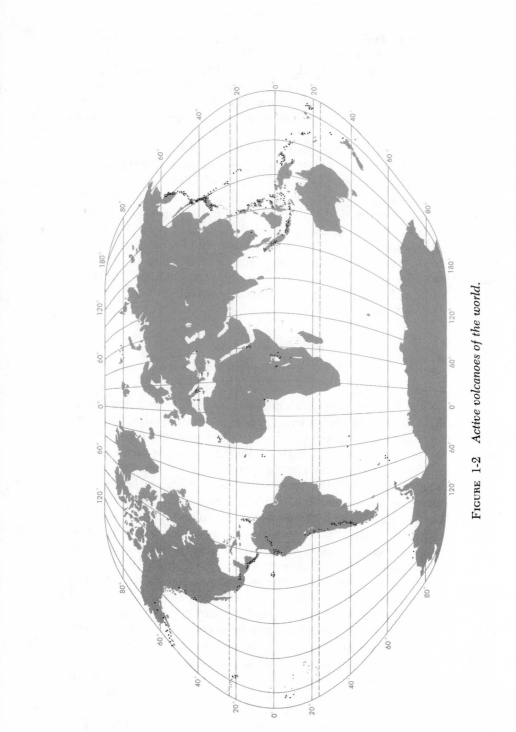

FIGURE 1-2 *Active volcanoes of the world.*

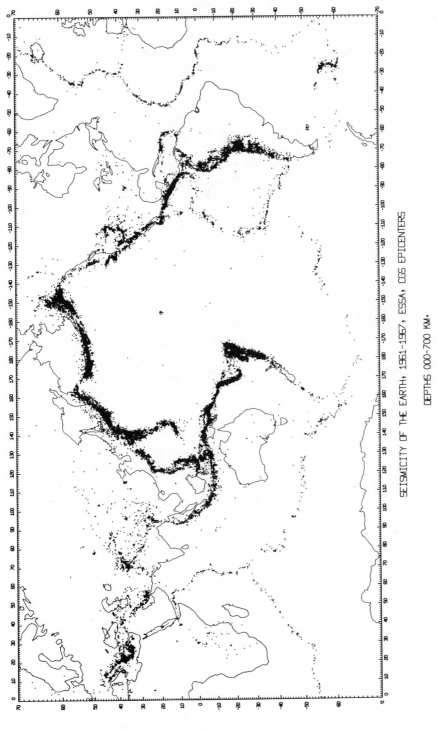

SEISMICITY OF THE EARTH, 1961-1967, ESSA, CGS EPICENTERS

DEPTHS 000-700 KM.

FIGURE 1-3 The locations of earthquakes in the years 1961–1967 compiled by the U.S. Coast and Geodetic Survey. From Barazangi and Dorman, 1961, Bull. Seis. Soc. Am., v. 59, p. 369.

Africa, Western North America, in Turkey, and in the Mediterranean on the southern margin of the Alpine mountain belt. A few centers of activity are not close to any of the topographic features that we have been discussing, but these represent just a small fraction of the active volcanoes. The great majority are closely associated with ocean trenches and to a lesser extent with ocean ridges and mountainous country.

The distribution of earthquakes shows a similar pattern (Fig. 1–3). The girdle of fire around the Pacific Ocean is again obvious and the greatest number of earthquakes occur adjacent to the deep trenches. But in Fig. 1–3 the ocean ridges are prominent as well. A narrow band of earthquakes marks the crests of the ridges in all of the world's oceans.

On land the pattern of earthquakes is more diffuse. A broad band of activity extends from Southern Europe, through the Near East, to Central Asia, coinciding with the Alpine-Himalayan Mountain chains. The North American Cordillera is similarly marked, as are the high plateaus of Eastern Africa. The failure of the earthquakes in these regions to fall on sharply defined lines is characteristic of the continents.

Thus, the distribution of volcanoes and earthquakes attracts our attention to those features of the Earth mentioned previously: mountains, ocean ridges, and ocean trenches. Of these, only the mountains are exposed above the sea for our direct observation and study. Consequently, they are known better and in more detail than the features beneath the sea. We shall proceed to an investigation of mountainous regions and defer until later the discussion of the oceanic features, which we can observe only indirectly.

Geologic structures

This chapter is concerned with the geological structures that we observe in land areas. We have reason to believe that these structures are more complex than their submarine counterparts and also that a more complete record of geologic history is preserved on land than is present beneath any of the world's oceans. The oldest continental rocks are about 3.4 billion years old, whereas rocks more than 200 million years old are rare or absent in the oceans.

Continental rocks are commonly observed to be deformed. We begin by defining two terms that are useful in discussing both the deformations and the process of mountain building. A *tectonic* process is one that produces deformations. An *orogeny* is the set of tectonic processes that eventually results in a mountain chain. For example, the tectonic processes that gave rise to the Alps can collectively be described as the Alpine orogeny. Belts of deformed rocks are termed *orogenic*, or *mobile*, belts whether they are presently mountainous or not. Such belts may be the last remnants of mountain ranges that were eroded away long ago.

The mobile belts surround areas called *cratons* or *shields*, which have been geologically inactive for a period of several hundred million years. Only slow erosion has affected the shields since their last deformation, and the oldest known rocks are exposed at the Earth's surface in such areas. The distinction between shields and mobile belts is relative rather than absolute however, for remnants of ancient mobile belts can be recognized within the confines of shields. Furthermore, many mobile belts have been inactive for hundreds of millions of years. They differ from the shields they surround by having been active more recently, but it is the relative rather than absolute age of the activity that serves as the criterion for distinction.

Faults and Folds

When deformed, rocks may either break or bend, producing faults and folds. Faults are fractures in the Earth across which relative motion has taken place, and the movement of the fault can be categorized further by its direction measured in the fault surface. Generally, there is a horizontal component to the motion and a component at right angles (Fig. 2–1). Faults with dominantly horizontal motion are termed *transcurrent faults*. When the motion is mainly in the direction at right angles, faults are classified as

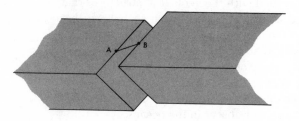

FIGURE 2-1 Generalized *motion on a fault. Points A and B, on opposite sides of the fault, were in contact prior to the faulting.*

normal or as *reverse faults*. The difference between the last two is illustrated in Fig. 2–2, which indicates that normal faulting involves extension, whereas reverse faulting involves shortening. Normal faults are characteristically steep, and steep reverse faults are common. But a special class of reverse faults is inclined to the horizontal by less than 30°. These are termed *thrust* or *overthrust faults*. The relative movement on great overthrust faults can be enormous, totalling many tens of kilometers.

When rocks are bent rather than broken, folds are formed (Fig. 2–3). The folds shown in the figure are obvious because the rock has a layered

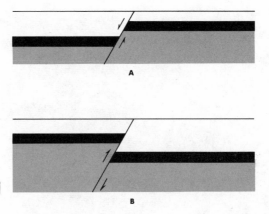

FIGURE 2-2 (A) Normal fault, (B) Reverse fault.

appearance that is due to *bedding*. These rocks were originally sediments, laid down in quiet water and later solidified into rock. Slight changes in conditions during deposition—changes in the velocity of currents supplying sediment, for example—affect the appearance of the resulting layers of rock and thereby make bedding visible. It is very important to realize that in most cases bedding develops in an essentially horizontal position. Sedimentary rocks that have not been deformed have horizontal stratification, as shown in Fig. 2–4. The striking contrast between Figs. 2–3 and 2–4 is due to the later folding of originally horizontal beds.

Nappes

A fold lying on its side, as in Fig. 2–5, is said to be *recumbent*. Recumbent folds may have very large horizontal dimensions, and in that case both they and large overthrust faults are collectively termed *nappes*. A nappe can be either a large recumbent fold or a large overthrust fault. Characteristically, nappes occur in groups, piled on top of each other to produce an extremely complicated overall structure.

Large nappes were first recognized in the Alps, and painstaking work by five generations of geologists has been required to unravel the complex picture. The results are shown as *structure sections* in Figs. 2–6 and 2–7. If we imagine a deep vertical trench cut through the mountains, a structure section shows the types of rock that would be exposed in its walls. *Vertical exaggeration* is commonly used in structure sections to show the relationships more clearly. If the vertical scale is 10 times the horizontal, the vertical exaggeration is 10:1.

Figure 2–6 shows a section through the northern Alps in the Glarus district of Switzerland. Two major nappes are shown in the figure; in this part of the Alps they take the form of thrust faults. Figure 2–7 shows the structure in the Simplon region, south of the area shown in Fig. 2–6 and near the Swiss-Italian border. Here the nappes take the form of gigantic recumbent folds.

The noteworthy features of Figs. 2–6 and 2–7 are the intensity of the deformation and the scale of movement. It is as if the rocks had been caught between the jaws of a giant vise and had responded by squirting northward like toothpaste from a tube. Masses of rock have traveled many tens of kilometers from their original positions. The rocks of the nappes are wildly contorted, but this feature cannot be shown on small-scale figures.

The intense compressive deformation that the Alps have suffered implies that the mountain chain, which is now about 150 km wide, was originally much wider. It is hard to estimate the initial width because of the complexity of the structure, but modern estimates range from 300 to over 600 km; that is,

FIGURE 2-4 *Well-bedded, undeformed sedimentary rocks exposed in the Goose-necks of the San Juan River, southeastern Utah. (Photograph by author).*

the opposite sides of the present Alps have approached each other by 150–450 km.

Two major problems are raised by structures of the Alpine type. How can nappes travel tens of kilometers over their substrata, and how is a crustal shortening of a few hundred kilometers, localized in a single mountain chain, accommodated by the earth as a whole? We shall proceed at once to a discussion of the first of these questions, deferring the second to a later chapter in this book.

Whether a nappe be a recumbent fold or an overthrust sheet, it must overcome the resistance of friction if it is to slide over the mass of rock it now covers. If the coefficient of friction between the nappe and the underlying rock is similar to that observed in the laboratory for dry rocks sliding across each other, it is easy to show that the maximum possible length of a nappe is about 10 km. In the case of a longer nappe, friction would com-

FIGURE 2-5 *A recumbent fold in a banded iron formation, Fig Tree sequence, Barberton mountains, South Africa. (Photography by author).*

FIGURE 2-6 *Structure section through the Glarus district, Switzerland. No vertical exaggeration. Two great thrust faults, each of which has moved older rocks over younger ones, are shown. The upper thrust has been eroded away in the center of the section and only appears at the ends. Adapted from E. B. Bailey, 1968, Tectonic Essays, Oxford University Press.*

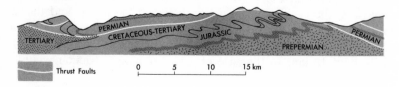

Thrust Faults

0   5   10   15 km

Geologic structures

14

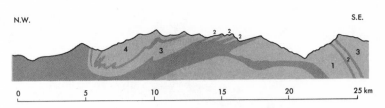

N.W.   S.E.

FIGURE 2-7 *Structure section along the route of the Simplon tunnel in Switzer-land. No vertical exaggeration. Four great nappes, each composed of a core of gneiss mantled by schist are distinguished by the numbers. Adapted from E. B. Bailey, 1968*, Tectonic Essays, *Oxford University Press.*

pletely inhibit motion, and the rock of the nappe would instead be crushed. Yet nappes are commonly observed to be much longer than 10 km.

A way out of this difficulty has been shown by M. K. Hubbert and W. W. Rubey. They suppose that near the lower surface of the nappe pore spaces in the rock are filled with water under a pressure that nearly equals the pressure due to the overburden of rock in the nappe. Under such conditions, the water partially "floats" the nappe and greatly reduces frictional resistance to motion. In principle, the friction would be reduced to zero if the water pressure equaled the pressure due to the rocky overburden. In practice, the pressure of pore water has been observed to be as large as 95 per cent of the pressure due to the overburden, which increases the theoretical maximum length of the nappe by a factor of 20 to about 200 km, a figure that is comparable with observed lengths.

The theory of Hubbert and Rubey provides a rational basis for understanding *gravity-sliding tectonics*. Masses of rock may slide long distances on gentle slopes if the coefficient of friction can be reduced sufficiently. The buoyancy provided by a high pressure of pore water gives rise to this reduction in friction, and many nappes and isolated blocks related to nappes may have moved into their present positions by this mechanism. Gravity-sliding is not restricted to consolidated rocks. Masses of soft sediment, buoyed by their interstitial fluid, can slump and slide down slopes on the sea floor, and certain puzzling geologic structures may have originated in this way.

Transcurrent Faults

Transcurrent faults present a problem that is similar to the problem raised by the amount of crustal shortening accompanying orogeny of the Alpine type. Major transcurrent faults have displacements measured in hundreds of kilometers. The best-known example is probably the San Andreas

Geologic structures

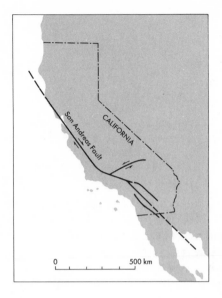
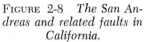

FIGURE 2-8 *The San An-dreas and related faults in California.*

Fault in California, Figs. 2–8 and 2–9, but many other faults with large horizontal displacements occur in lands bordering the Pacific. Another well-known transcurrent fault is the Great Glen Fault, which shifts the northern part of Scotland about 100 km westward relative to the southern part (Fig. 2–10). The problem raised by such faults is how they terminate.

FIGURE 2-9 *Aerial view of the trace of the San Andreas fault, Carrizo Plain, California. Note the offset of the sharp stream gulley left of center in the picture. (Photograph by John S. Shelton.)*

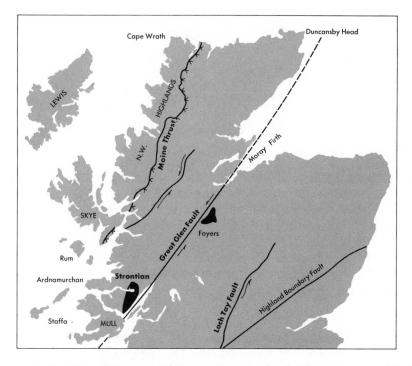

FIGURE 2-10   *Map of the Great Glen Fault in Scotland. The two masses labeled Strontian and Foyers represent what was once a single body of granite that was cut in two and separated by horizontal motion along the fault.*

It is not hard to imagine how a transcurrent fault terminates at depth. It presumably passes downwards into a region where, because of high temperature and pressure, it simply loses its identity in plastic low at that level. But this explanation does not provide a means by which the fault can end at the surface. A different explanation is offered in Fig. 2–11, which suggests a way in which folding in response to compressive forces might let the fault terminate. Such an explanation may be satisfactory if the motion is not too great, but the amount of folding near large faults like the San Andreas or Great Glen appears much too small to absorb the observed displacements on the faults. Both of these faults pass out to sea at their ends, so their terminations cannot be observed.

The total displacement on the San Andreas Fault is still a controversial figure, but it may be as large as 500 km. The time that the fault has been active is also in doubt, but it could be of the order of 100 million years. If the fault originated more recently, the figure for the total displacement

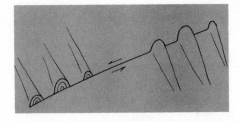

FIGURE 2-11 A way in which folding may accommodate transcurrent faulting and permit the fault to terminate.

would have to be revised downward. The velocity of relative motion along the fault comes out to roughly 0.5 cm/year on the average. The present rate of motion, measured by repeated surveys across the fault, is as large as 5 cm/year. This latter figure may well represent local values of extremely rapid motion that are not representative when averaged over the length of the fault or through extended periods of time.

These large figures for the rate and total amount of relative motion on the San Andreas Fault are not universally accepted. Some geologists contend that there is no evidence for total motion exceeding 50 km. The discrepancy originates in the ambiguous nature of the evidence used to infer the total displacement; neither the lower limit of 50 km nor the upper one of 50 km can be shown to be impossible. Current opinion favors the larger values of displacement, which makes it all the more difficult to understand how the San Andreas, and similar faults, terminate. We shall defer this question to a later chapter, as we did the problem of crustal shortening in the Alps.

Structures of Some Other Mountain Systems

The kinds of complexities shown by Alpine structure are not unique to that mountain chain, but they are not shown by all mountains. In New England the Appalachian mountains have Alpine structures, and the Himalayas probably do too, although little work has been done because the region is so inaccessible. But other mountain systems do not show characteristic Alpine features, and we shall discuss a few examples.

A region of classic geologic structure is the Valley and Ridge Province of the Appalachian Mountains. The province derives its name from its topography, which consists of long, even-crested ridges with valleys intervening (Fig. 2–12). The ridges are capped by hard sandstones that resist erosion, whereas the valleys are eroded out of softer strata. Thus, the structure of the sedimentary rocks controls the topography.

A structure section through the Appalachians in central Pennsylvania is

FIGURE 2-12 *Aerial view of the Valley and Ridge Province near Harrisburg, Pennsylvania. (Photograph by John S. Shelton.)*

shown in Fig. 2–13. The dominant structures are the great folds. The deformation is fairly intense; beds are occasionally rotated through the vertical into an inverted position. (No such strata are shown in Fig. 2–14, but they occur elsewhere in the Valley and Ridge Province.) A few thrust faults are present, but they are of minor importance in central Pennsylvania.

Further to the south, in southeastern Tennessee, the character of the deformation is different, as shown in Fig. 2–14. Here thrust faults predominate over folds. They are believed to flatten with depth, and they follow mechanically weak strata. Thus, they tend to be parallel to the bedding, which

FIGURE 2-13 *Structure section through the Valley and Ridge Province in central Pennsylvania. No vertical exaggeration. Precambrian (pЄ), Cambrian (Є), Ordovician (O), Silurian (S), and Devonian (D) strata are shown. Adapted from the Geologic Map of Pennsylvania, 1960, Commonwealth of Pennsylvania, Department of Internal Affairs.*

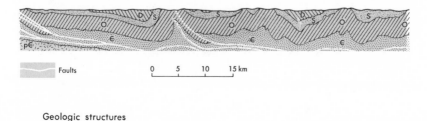

Faults   0   5   10   15 km

Geologic structures

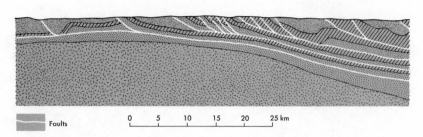

Faults

0   5   10   15   20   25 km

FIGURE 2-14   *Structure section through the Appalachians in southeastern Tennessee. No vertical exaggeration. Adapted from Rodgers, 1953, Kentucky Geological Survey Special Publication, No. 1.*

makes their recognition difficult since they produce no visible offsetting of strata. Nevertheless, evidence from oil and gas wells drilled in the southern Appalachians supports the interpretation given in Fig. 2–14. The displacements on the faults may amount to many kilometers.

The rocks of the Valley and Ridge Province have been folded and faulted to a marked degree, but the deformations they have undergone are not nearly as intense as those found in the Alps. The Valley and Ridge Province felt the effects of the last compressive phase of an organic cycle that began hundreds of millions of years earlier in other parts of the Appalachians and produced far more violent deformations elsewhere.

The high Andes in southern Peru are entirely different structurally than the Alps and Appalachians. Figure 2–15 is a structure section that runs from near the coast, through the city of Arequipa, and to the magnificent volcano, El Misti (Fig. 2–16). The profile crosses several large bodies of igneous rock, which crystallized from a melt at depths of many kilometers in the Earth. These bodies are separated by masses of older sedimentary and volcanic rocks that form the crests of a range of foothills near the middle of the section. The highest ground is underlain by Cenozoic lavas culminating in the 5800-meter summit of El Misti, which is still active. The high Andes owe their elevation mainly to the building of volcanoes, rather than

FIGURE 2-15   *Structure section through the Peruvian Andes. No vertical exaggeration. Large masses of igneous rocks are labeled ig, and the diagonally ruled rocks near El Misti are volcanic. Adapted from W. F. Jenks, 1948, Instituto Geologico del Peru, Bulletin No. 9.*

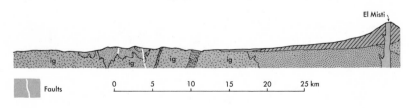

El Misti

ig   ig   ig   ig

Faults

0   5   10   15   20   25 km

Geologic structures

20

FIGURE 2-16 *View of the volcano El Misti, northeast of Arequipa, Peru. (Photograph by F. R. Boyd).*

to faulting and folding followed by erosion. The volcanoes achieve most impressive elevations. Together with the high plateau that surrounds them (the Altiplano), they form the second largest area higher than 4000 meters that is to be found on the Earth; only the Himalayas and the Tibetan plateau is larger.

A third structural cause of mountainous topography is exemplified by the ranges in the Great Basin of the southwestern United States (Fig. 2–17). The mountains owe their relief to normal faulting, which produces elevated mountainous blocks separated by basins that are filled with recent sediments. The term *fault-block mountains* is applied to such structures. Within the uplifted blocks the deformation may be as intense as in the Alps, but

FIGURE 2-17 *Structure section crossing Nevada at latitude 37° N. Vertical exaggeration 5:1. Adapted from Hamilton and Pakiser, 1965, U.S. Geological Survey, Miscellaneous Geologic Investigations, Map I-448.*

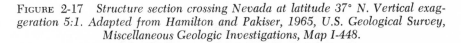

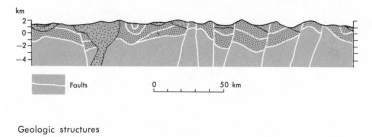

Geologic structures

these structures formed by compression predate the movements that caused the present topography. The dominance of regional extension over compression is demonstrated by the prevalence of normal faults bounding the ranges.

In the Appalachians, Triassic sedimentary rocks occupy structural positions equivalent to the sediments in the intermontane basins of the Basin and Range. These are the youngest rocks preserved in the Appalachians, and presumably they represent the remnants of basins, similar to those in the Basin and Range, that resulted from tensional forces following the compression that produced the folds in the Valley and Ridge Province.

As a final example of mountains, we consider the Barberton mountains

FIGURE 2-18 *Schematic structure sections illustrating the evolution of the Barberton mountains. (A) Deposition of volcanic and sedimentary rocks, accompanied by intrusion of molten rock. (B) Folding and faulting of the rocks accompanied by the intrusion of great masses of granite. (C) Differential erosion leaves the sedimentary and volcanic rocks as a mountain land. (Faults are shown in white.)*

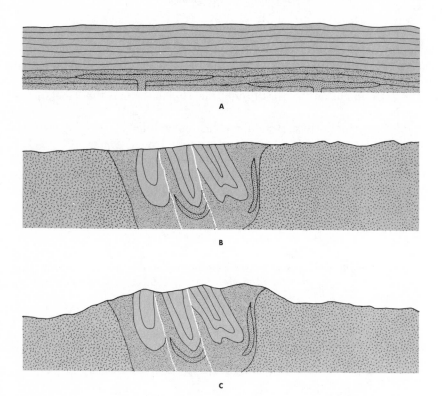

of South Africa. The Barbertons owe their mountainous topography to the fact that the rocks of which they are composed are more resistant to erosion than are their surroundings; as a result, they stand up above the surrounding plains.

The rocks in the Barberton mountains are among the oldest on Earth. They owe their survival to their structure, which protected them from the erosion that has destroyed rocks of comparable age in the surrounding region. They occur as a series of "keels" that are folded and faulted downwards in such a way as to be incorporated into a sea of younger granites produced by solidification of molten rock that invaded the area subsequent to the formation of the Barberton rocks. The development of the Barberton mountains is shown schematically in Fig. 2–18.

Rift Valleys

Most valleys are the products of erosion by streams, but some are of structural origin. The term *rift valley* was applied by J. W. Gregory to elongate valleys with steep parallel sides which cut through the high plateaus of eastern Africa (Fig. 2–19). Structurally these valleys are *grabens*, which are blocks dropped down between parallel faults (Fig. 2–20a). The

FIGURE 2-19 *The rift valleys of Eastern Africa.*

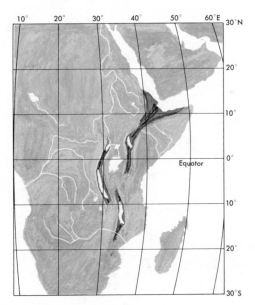

walls of the rift valleys are the escarpments formed by the bounding faults.

Rift valleys have been recognized in many parts of the world. Examples are the Red Sea-Gulf of Aden Depression, the valley of the river Rhine, a deep depression in central Asia partly occupied by Lake Baikal, the valley of the Dead Sea, and the rift valleys of Africa. These valleys are often very deep. The bottom of Lake Baikal, the deepest lake in the world, is about 1 kilometer below sea level, and the surface of the Dead Sea is about 400 meters below sea level. Rift valley floors are commonly covered with thousands of feet of volcanic rocks and by sediments deposited in the lakes that characteristically occupy these tectonic depressions. Hence the depth of the valley may be very much less than the total amount of vertical movement along the boundary faults, which can be demonstrated to exceed 7000 meters at one point on the African Rift.

Rift valleys are geologically active features. Volcanoes and earthquakes occur within and adjacent to the rifts, although they are not restricted to the valleys themselves. Hot springs and steam vents are common in the valleys. Movement along the swarms of faults that cut the valley floors has been very recent, as is shown by the fact that young sediments are displaced and erosion has not had time to modify the scarps formed by the faulting.

A glance at Fig. 2–19 shows that rift valleys are major geologic features, but so far our understanding of them is incomplete. Early observers suggested that the Earth's crust was pulled apart, allowing a keystone-shaped block to drop downwards along normal faults as shown in Fig. 2–20a. But the East African rifts cut a plateau that was being uplifted contemporaneously with the formation of the valley. Furthermore the amount of uplift is greatest at the margins of the rift valleys. Simply pulling the crust apart does not explain the regional uplift that is closely associated with the rifting. These objections led to a rival theory, illustrated in Fig. 2–20b. The

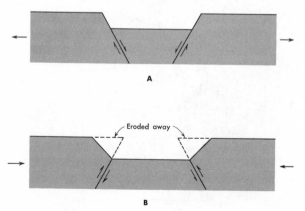

FIGURE 2-20 *The origin of rift valleys (A) by tension and (B) by compression.*

FIGURE 2-21 *Open fissures in the floor of the rift valley on the flanks of Mt. Fantale near Awash Station, Ethiopia. (Photograph by author.)*

crust is assumed to be under compression and the plateaus ride up over the rift valley on reverse faults. The valley is thereby forced downward, uplifting the surrounding plateau. But this theory also encounters difficulties. It is not at all clear that an area of uplift many times as large as the area of the valleys could be created in this way. Where the dips of the faults can be determined, they are either normal or the fault plane is vertical. (In the latter case neither tension nor compression can explain the faulting.) A few examples of open fissures have been found in the valley floors (Fig. 2–21). Thus the geological evidence suggests that the crust is in a state of tension rather than compression, favoring the hypothesis of Fig. 2–20a. It seems that the rift valleys cannot be explained by purely horizontal forces. Vertical forces, the nature and origin of which are still obscure, must also be considered.

Another question about rifting is how far the process can go. Does the Red Sea, which divides two continents, represent an advanced stage of a process which began like, say, the East African rifts? Will Africa and the other continents where rift valleys are recognized eventually be broken up? Can rift valleys become seas the size of the Mediterranean? Can they become oceans? We cannot answer these questions at present.

Geologic structures

3

The Earth's magnetic field

In this chapter we shall describe the Earth's magnetic field, discuss the magnetism of rocks, and consider some surprisingly far-reaching implications about the history and development of the Earth's crust that have resulted from the study of geomagnetism. We shall say nothing about extraterrestrial influences on the Earth's field, and we can only briefly introduce the subject of its origin.

A magnetic field is a *vector* field; that is, it has magnitude and direction. A magnetized needle suspended so that it can freely assume any orientation whatsoever will align itself in the direction of the local field. The direction of the needle, plus the intensity of the field at the same point, completely characterizes the field. Magnetic fields are conventionally represented by *lines of force*, which are lines that point everywhere in the direction of the field. The intensity of the field is represented by the distance separating adjacent lines of force. Where they are close together the field is strong; where far apart it is weak. The Earth's field is nearly that of a magnetic dipole. The lines of force for a dipole field are shown in Fig. 3-1a, superimposed on the Earth.

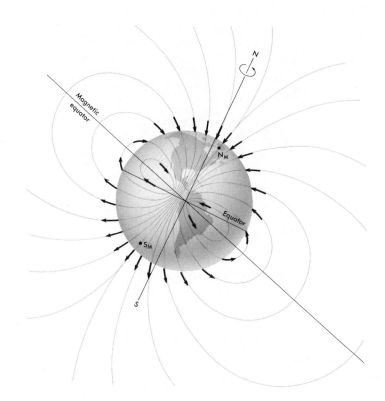

FIGURE 3-1 (A) The magnetic field of a dipole. A magnetized needle would point along the lines of force (white).

There are two points where the lines of force are vertical. These are the magnetic poles, which are 180° apart if the dipole is centered in the Earth. The Earth's magnetic poles are separated from the geographic poles by about 18°. The North Pole is in the islands north of Canada; the South Pole, in Antarctica south of Tasmania. The intensity of the field is approximately 0.6 gauss at the magnetic poles and 0.3 gauss at the magnetic equator.

The Earth's field actually departs somewhat from a dipole. The departure is shown in Fig. 3–2, which is a map of the difference in vertical intensity between the Earth's field and the dipole field that most nearly matches it. We note that a date is affixed to Fig. 3–2. This is because the field changes with time, a phenomenon known as the *secular variation*. A striking part of the secular variation is the *westward drift*, which is a westward movement of the characteristic humps and hollows of the nondipole field as well

FIGURE 3-1 *(B) The strength of a dipole field (dashed) compared to the Earth's field in 1945. Lines of constant field strength are shown in gauss.*

as the magnetic poles themselves. The motion relative to the Earth's surface is such that the features of the field would circle the Earth in a few thousand years. The westward drift is not the whole of the secular variation however; the features of the field change their shapes as well as their locations. Finally, the secular variation is a worldwide phenomenon.

The Dynamo Theory

The modern theory of the Earth's magnetic field is attributed mainly to E. C. Bullard and W. M. Elsasser. It is based on the following deductions. The worldwide nature of the secular variation suggests that the field originates at great depth, for otherwise a close correlation between local shallow phenomena on a worldwide basis would be required. It is hard to imagine how such correlations would be brought about. The time scale of the secular variation, of the order of 1000 years, is much too short to be connected with geologic events affecting the solid part of the Earth. These have a time scale of millions of years. On the other hand, the secular variation is too slow to be attributed to the atmosphere or oceans. The magnetic poles are close enough to the geographic poles to suggest that the Earth's rotation may have something to do with its magnetic field.

There is evidence, discussed in a later chapter, that the outer part of the

Earth's core is a liquid metal. As such, it is an electrically conducting fluid. A fluid of this type, when in motion, can interact with a magnetic field, and the field, in turn, influences the motion of the fluid. The study of these interactions is called *magnetohydrodynamics* in Europe; in North America the briefer term *hydromagnetics* is sometimes used. Regardless of the name, it is a complex field, but enough progress has been made to indicate that the *regenerative dynamo* of Bullard and Elsasser is at least a theoretical possibility.

According to the dynamo theory, the Earth starts with no magnetic field of its own. But weak magnetic fields are always present in the Galaxy, and if one is present when fluid motions are occurring in the core, the field will influence the motions. Under suitable circumstances the pattern of motions will create a magnetic field of its own. A weak galactic field is regenerated by dynamo action in the Earth's core, giving rise to the much stronger magnetic field of the Earth.

It is known that the two ingredients, a weak initial field and fluid motions, are insufficient by themselves to create dynamo action. But fluid motions in the Earth are also influenced by the Earth's rotation. Rotation provides a vital third ingredient that makes a regenerative dynamo possible. A reason why Venus has no magnetic field may be the very slow rotation of that planet.

FIGURE 3-2   *The nondipole vertical magnetic field. Shown are lines of constant difference between the vertical component of the Earth's field in 1945, and the vertical component of the best-fitting dipole field. Units are gauss. Data from* Bullard et al., 1950, Phil. Trans. Roy. Soc. London, v. 243A, p. 67.

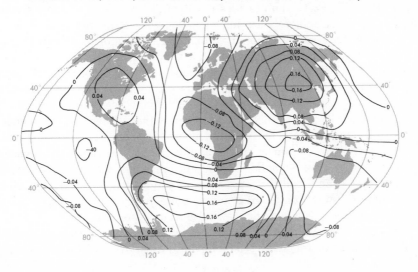

The Earth's magnetic field

The theory of a regenerative dynamo is very complicated, and there is little more that we can say about it here. Although the dynamo theory has never been proven correct, it is certainly possible, and it is freer from objection than any other theory proposed to date. The vital role played by rotation in the theory suggests that the magnetic poles should of necessity be close to the geographic poles. It is easy to find reasons why the locations of the two poles should not coincide exactly, but it would be hard to account for magnetic poles at much lower latitudes than they occupy today.

The fluid motions that produce the dynamo of Bullard and Elsasser are somewhat unstable, and resulting irregularities in the motion produce the secular variation. Instability can cause the sense of the field to reverse in a relatively short time, so that the north magnetic pole, as conventionally defined, appears near the south geographic pole. Such *reversals* of the field, occurring as a result of the instability of the dynamo, are believed to have taken place many times in the Earth's history.

Rock Magnetism

A few minerals are magnetic, and the rocks in which they occur can themselves become magnetized. It is possible to measure both the intensity and the direction of magnetization of the rock, and the latter quantity enables us to determine the direction of the Earth's magnetic field in the past. Magnetic minerals contain atoms of certain elements, of which iron is the only naturally important example, that behave as tiny magnets. In certain crystals the atomic magnets are oriented parallel to each other and so the material is magnetic. The important magnetic minerals are magnetite, Fe_3O_4, hematite, Fe_2O_3, and to a lesser extent ilmenite, $FeTiO_3$, and pyrrhotite, $Fe_{1-x}S$. These minerals are all *ferrimagnetic*, as distinct from a *ferromagnetic* material such as metallic iron. The distinction is shown in Fig. 3–3. In ferromagnetic material all of the atomic magnets are aligned in a common direction, whereas in ferrimagnetic substances some of the magnets point in one direction and the others point in the opposite direction. But either the number of magnets, or their strengths, are unequal in the two directions, so that a net magnetism remains. When a ferrimagnetic crystal is heated above a fairly well-defined temperature, called the *Curie point*, the common alignment of the atomic magnets is destroyed, and the crystal becomes *paramagnetic*. The atomic magnets are now oriented at random (Fig. 3–3c), and the magnetism of the crystal is weaker by orders of magnitude. Most iron-bearing minerals are paramagnetic even at room temperature, which is why their magnetism is unimportant.

The manner in which a rock acquires its magnetization depends on the type of rock. Lavas crystallize at temperatures above the Curie points of

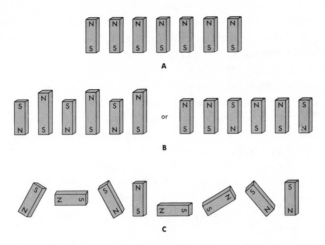

FIGURE 3-3 *Configuration of atomic magnets in (A) ferromagnetic materials, (B) ferrimagnetic materials, and (C) paramagnetic materials.*

their magnetic minerals. As they cool through the Curie points there is a tendency for the minerals to acquire magnetization in the direction of the field that is present at the time. It is theoretically possible for a rock to become magnetized in a direction opposite to the ambient field, but this phenomenon, termed *self-reversal*, appears to be rare in nature. The magnetism acquired by cooling through the Curie point is termed *thermoremanent magnetization*. Sedimentary rocks, which have never been heated, can also be magnetized. This is called *isothermal remanent magnetization*, and it comes about in two main ways. If some of the grains in the rock are already magnetized, during sedimentation they will have a slight tendency to line up with the existing field, just as a compass needle does. Even a small amount of preferred orientation produced in this way will give the rock measurable magnetization. Alternatively, magnetic minerals may be chemically precipitated in the pores of the sediment, as in the case of red sandstones cemented with hematite. These minerals tend to grow in such a way that their magnetism is parallel to the local field at the time. No mechanism of self-reversal has been suggested for sediments.

Paleomagnetism

The study of the magnetism of rocks and, in particular, of the directions in which rocks are magnetized can tell us the rock's position on the surface of the Earth at the time it was magnetized. As might be suspected, certain assumptions must be made to enable us to extract such information from magnetic data. Some of these assumptions can be tested and their validity established, but others lack experimental confirmation although they are theoretically plausible.

A primary requirement of paleomagnetism is that the magnetism acquired by a rock when it is formed must be in the direction of the field at the time. The rock must retain its original direction of magnetization even if it is moved and as a consequence exposed to the Earth's field in a different direction for hundreds of millions of years. This second property, known as *stability*, is not shown by all rocks, but reliable tests for stability have been devised. Since it is subject to verification, stability is not a separate assumption of paleomagnetism; we need only assume that the rock was magnetized in the direction of the field prevailing at the time of its formation.

A second assumption of paleomagnetism is that the rocks were magnetized in a dipole field. A test of this assumption would be to measure magnetization directions in rocks of the same age from several widely separated sites. But this test requires that we know the past locations of the sites relative to one another, and this we do not know. The best that can be done is to study rocks of the same age from as far apart as possible on the same continent. Where such tests have been made, the results are consistent with a dipole field. However, the discrimination is not as good as would be provided by a set of worldwide samples, the relative positions of which were known. There are also theoretical justifications for assuming a dipole field, but these follow from the dynamo theory, which is not yet fully worked out. Other forms of magnetic field cannot be excluded. The assumption of a dipole field is consistent with the field today, is consistent with what little we know of the field in the past, and has the virtue of being the simplest assumption we can make. It will not be abandoned until evidence is presented to show clearly that it is wrong.

As we shall see, the assumption of a dipole field leads to the conclusion that some paleomagnetic sites have moved relative to each other. If we abandon the dipole assumption and assume instead that these motions have some particular pattern (we could, for example, assume no motion at all), we then could deduce the geometry of the magnetic field that would be consistent with the magnetic observations and the assumed locations of the sites. But this new set of assumptions would be no more plausible than assuming a dipole field, nor would it rest on as sound an observational basis. In many cases, wildly irregular shapes of the lines of force would be required, and we have no guarantee that a mechanism exists for producing such a field.

The final assumption of paleomagnetism is that the magnetic and geographic poles coincide. The south magnetic pole might fall at the north geographic pole as the result of a magnetic reversal, but the difference in latitude between the poles is taken to be zero. This assumption might seem surprising in view of the fact that the magnetic poles are currently at latitude 72° rather than 90°, but some of the earliest paleomagnetic work provided experimental support. Study of a series of rocks from the western United States showed that the directions of magnetization pointed to true north more frequently than any other direction, including the present direction of the field. Data are shown in Fig. 3–4; the results were obtained on sedimentary rocks, and each sample contains strata that have a range in age of tens or even hundreds of thousands of years. This time is much longer than the time scale of the secular variation, and the magnetic poles have had time to circle the geographic poles several times. The average position of the magnetic poles does not differ measurably from the geographic poles. A second consequence of the range in age of the sample is that the nondipole components of the field tend to average to zero. Volcanic rocks cool through their Curie points quickly however, and determining the magnetism of a single sample will yield the direction of the instantaneous magnetic pole at the time. The "average pole," which coincides with the geographic pole, can be determined by sampling a number of lava flows of slightly different ages and averaging their directions of magnetization.

Observational confirmation of the coincidence between the magnetic and geographic poles comes only from very recent rocks. Our justification for

FIGURE 3-4 *Directions of horizontal components of magnetization of sedimentary rocks from the western United States. Data from Torreson, Murphy, and Graham, 1949,* Jour. Geophys., Res., v. 54, p. 111.

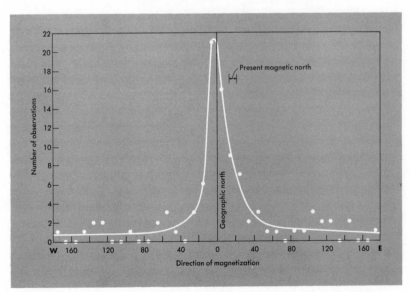

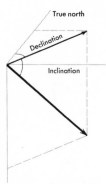

FIGURE 3-5 *Declination and inclination of a magnetic direction.*

assuming it throughout geologic time is largely derived from the important role that rotation plays in the dynamo theory. But many of the most interesting paleomagnetic results do not depend on this particular assumption. We could abandon all pretense of a geographic coordinate system and base our results solely on magnetic coordinates. Any *relative* motions of paleomagnetic sites would be unaffected. Only motions of the Earth's entire crust, acting as a rigid body, have been misinterpreted if the magnetic and geographic poles have not always coincided in the past.

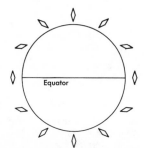

FIGURE 3-6 *Inclination as a function of latitude for a centered axial dipole field.*

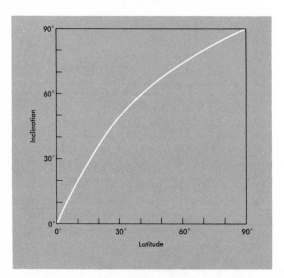

The direction of magnetization of a rock is given in terms of the *inclination* and *declination*. The former is the angle that the direction of magnetization makes with the horizontal, and the latter is the true bearing of the horizontal component of the magnetization (Fig. 3–5). In the case of a centered axial dipole field, with coincident geographic and magnetic poles, the inclination depends only on latitude and not on longitude, and the declination is always zero; that is, the horizontal component of the field points towards the poles along lines of longitude. The relation between inclination and latitude is

$$\tan I = 2 \tan \theta,$$

where I is the magnetic inclination and θ is the latitude. Inclination is shown as a function of latitude in Fig. 3–6. Nonzero declination, which is

FIGURE 3-7 *Paleolatitudes of Cincinnati, Ohio and Paris, France. Longitudes have no significance and it is assumed that drift was along a great circle. Arrows indicate the orientations of the paleomeridians, and numbers beside the points indicate ages in millions of years. Data from E. Irving, 1964,* Paleomagnetism, John Wiley & Sons.

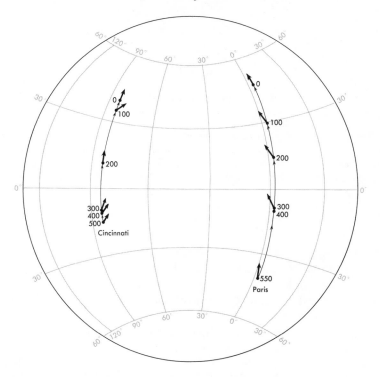

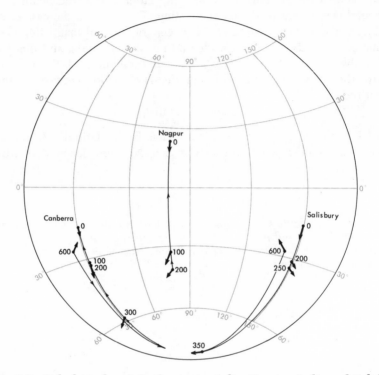

FIGURE 3-8 *Paleolatitudes of Canberra, Australia, Nagpur, India and Salisbury, Rhodesia. Longitudes have no significance and have been arbitrarily altered for clarity. It is assumed that drift was along a great circle. The abrupt changes in drift direction shown by Africa and Australia simply indicate that the continent drifted over the pole. Arrows indicate the orientations of the paleomeridians and numbers beside the points indicate ages in millions of years. Data from E. Irving, 1964,* Paleomagnetism, *John Wiley & Sons, and from McElhinny et al.,* Rev. Geophys., *v. 6, p. 201.*

commonly observed, shows that the site has been rotated relative to the poles. It represents the angle that the paleomeridians make with the present meridians of longitude.

Paleomagnetic data for five continents are shown in Figs. 3–7 and 3–8. The significant data are the paleolatitudes as plotted and the paleomeridians, shown as arrows. Longitudes have no significance, and displacements in longitude are done simply for clarity. It is impossible to tell unambiguously whether a paleomagnetic site is in the northern or southern hemisphere because of magnetic reversals. The hemispheres are chosen to minimize con-

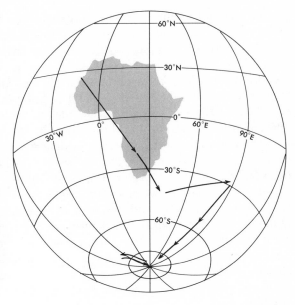

FIGURE 3-9 *Alternative way of displaying the paleomagnetic data for Africa. The continent is regarded as fixed and the pole is then considered to wander. Arrows indicate path of the pole from 600 million years ago to the present. Data from McElhinny et al., 1968, Rev. Geophys., v. 6, p. 201.*

tinental motion. The reversals in direction of motion shown by Africa and Australia in Fig. 3–7 are apparent rather than real. If a point moves over either pole, its direction of motion in latitude reverses and its longitude changes by 180°. Such large changes in longitude were not considered in constructing the figure. An alternative plot of the paleomagnetic data for Africa is shown in Fig. 3–9.

The reference localities shown in Figs. 3–7, 3–8, and 3–9 have undergone large displacements relative both to the poles and to each other. Such displacements are termed *continental drift.* The drift curves for Eurasia and North America, as shown in Fig. 3–7, are rather similar. Neither continent moved much between 300 million and 400 million years ago; at other times the relative motions were small but appreciable. The southern continents, however, underwent large relative displacements. In the early Paleozoic Era they drifted towards the south pole, whereas the northern continents moved northward. Peninsular India moved northward by about 65° in the last 200 million years according to Fig. 3–8, and Fig. 3–7 shows that Eurasia moved only 35° in the same period. This is evidence that peninsular India has not always been part of Asia. Is it mere coincidence that the junction between these formerly separate land masses is now covered by the highest mountains on Earth, the Himalayas?

Other Evidence for Continental Drift

This section represents somewhat a digression from the subject of magnetism, but we include it here to show that the evidence on which continental drift is based is far broader than the magnetic data that we have been discussing. The idea of continental drift can be found in, or read into, the writings of Francis Bacon in 1620. He was impressed by the fact that the eastern coast of the Americas could be fitted rather neatly into the western coasts of Europe and Africa. The idea of continental drift was elevated into real prominence early in this century by the German meteorologist Alfred Wegener. His most impressive evidence came from indications of past climates, particularly in the Carboniferous and Permian Periods. In those times Africa, Australia, Antarctica, South America, and peninsular India underwent glaciation. At much the same time coal was being formed in North America, Europe, and Asia, and desert conditions prevailed elsewhere in the north. Coal deposits and deserts are indicative of warm climates that occur near the equator today. Glacial conditions are found predominantly in polar regions. The ancient climates cannot be made to agree with the present climatic pattern if no relative motion of the continents is allowed. Any choice of pole position leaves at least one of the glaciated regions in low latitudes.

Wegener showed that by grouping all of the continents into a single land mass in Permocarboniferous times, one could choose a pole near South Africa and have all of the glaciated regions in its vicinity. This arrangement of the continents would also leave the presumed tropical regions near the equator. Geologists have modified Wegener's ideas to make them conform more closely to their evidence. The four southern continents and peninsular India have similar geologic histories that are in sharp contrast with the histories of the northern continents, particularly in the Paleozoic Era. In the north, where the classic geologic successions were established, rocks of Precambrian age form the crystalline basement upon which a complete sequence of Paleozoic rocks was deposited. In the south the basement includes rocks ranging up to Silurian age. Unmetamorphosed sediments older than Carboniferous are rare. A grouping of the continents as they might have been near the end of the Paleozoic is shown in Fig. 3–10. The data on which the figure is based includes paleomagnetism. The continents are grouped into a northern landmass called Laurasia and a southern one called Gondwanaland. A very conspicuous feature of the figure is the narrowness of the Atlantic Ocean. If this reconstruction is correct, the Atlantic must be relatively new, formed within the last 300 million years.

No single line of evidence taken by itself can be considered definitive proof of continental drift. Paleomagnetic evidence is subject to the assumptions discussed in the last section. At present, we do not understand climate

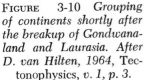

FIGURE 3-10 *Grouping of continents shortly after the breakup of Gondwanaland and Laurasia. After D. van Hilten, 1964, Tectonophysics, v. 1, p. 3.*

in a quantitative sense, and there could be conditions under which widespread glaciation can occur near the equator. The geological evidence might be coincidental. But when all of these lines of evidence are considered together, the fact that they suggest drift in the same direction and of roughly the same magnitude cannot be overlooked. Large-scale relative motions at the surface of the Earth would seem to be an important part of tectonism.

The Vine-Matthews Hypothesis

The Vine-Matthews hypothesis was proposed as the result of extensive studies of magnetic anomalies. Before describing it we must define what is meant by an *anomaly* in geophysics. Many large-scale features of the Earth, its gravitational and magnetic fields for example, are measured with such high accuracy that five to seven figures are required to record the number. Furthermore, such fields characteristically show smooth changes on a global scale, upon which local irregularities are superimposed. It is often convenient to subtract these large-scale features in order to make the smaller local features more prominent. In the process, the number of significant figures that need be carried is reduced to four or less. An anomaly is defined as the difference between a measured quantity and some reference standard that is usually chosen to represent its broad, worldwide variations as closely

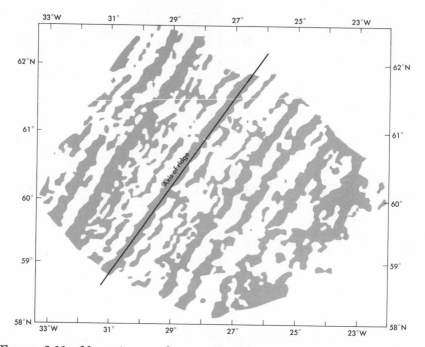

FIGURE 3-11 *Magnetic anomalies near the crest of the Mid-Atlantic Ridge. Gray areas represent positive anomalies, white areas negative ones. After Heirtzler, Le Pichon, and Baron, 1966, Deep-Sea Res., v. 31, p. 427.*

as possible. In the case of the vertical intensity of the magnetic field, we might choose the sum of the vertical component of the best-fitting dipole field and the nondipole components shown in Fig. 3–2 as our reference standard. Subtracting the vertical intensity predicted by this combination from the observed vertical intensity leaves the magnetic anomaly at the point of observation.

As magnetic observations in the deep sea became more numerous, a striking pattern emerged. Belts of alternating positive and negative anomalies are oriented parallel to prominent ocean ridges such as the Mid-Atlantic Ridge and the East Pacific Rise. Figures 3–11 and 3–12, in which regions of positive anomaly are dark and negative anomalies are light, illustrate the patterns observed. Further features of these data are brought out in Fig. 3–13. The middle line shows a magnetic profile across the East Pacific Rise, and the top line shows the same profile reversed so that its east end, rather than its west end, is at the left-hand side. These two lines illustrate the high degree of symmetry of the magnetic patterns about the ridge crest. The pro-

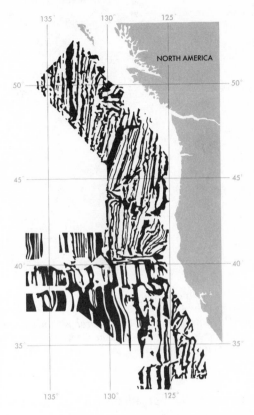

FIGURE 3-12 Magnetic *anomalies off the West Coast of North America. Black areas represent positive anomalies, white areas negative ones. After A. D. Raff, 1966, Jour. Geophys. Res., v. 71, p. 2361.*

file to the east of the ridge is almost a perfect mirror image of the profile to the west. The third line in Fig. 3–13 shows a magnetic profile across the Mid-Atlantic Ridge. The horizontal scale, which differs from the first two lines, has been chosen to make the principal features of the profiles correspond as closely as possible. When this is done, the similarity between all three profiles is striking.

What could produce such elongated anomalies, symmetrical about the ridge crests, and similar from ocean to ocean except for their scales of distance? F. J. Vine and D. H. Matthews put forth the bold hypothesis that the anomalies are due to thermoremanent magnetization in basaltic rocks in the crust underlying the sea floor. Belts of positive anomalies would be due to rocks magnetized in the direction of the Earth's field and reinforcing it; negative anomalies would occur where the rocks are magnetized in the opposite direction and oppose the field. And what could cause such belts of opposite magnetization? The only plausible explanation that comes to mind

The Earth's magnetic field

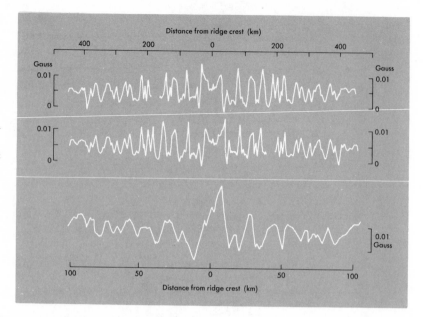

FIGURE 3-13 *Magnetic anomaly curves from the East Pacific Rise and the Mid-Atlantic Ridge. The upper horizontal scale is for the two top curves and the lower scale is for the bottom curve. The poorer resolution shown by the bottom curve is due to the shorter horizontal distance actually covered. See text for further description. Data from Pitman and Heirtzler, 1966, Science, v. 154, p. 1164.*

immediately, and the one advocated by Vine and Matthews, is reversals of the Earth's magnetic field.

Reversals of the magnetic field are known to have occurred, and the times of some of the most recent reversals have been established by careful study of well-dated sequences of rocks. A promising method of dating young rocks has emerged from this work and is discussed in *Geologic Time,* by D. L. Eicher, and in *Oceans,* by K. K. Turekian. But magnetic reversals are worldwide phenomena that have occurred at definite times in the past, not local events affecting belts parallel to ocean ridges. The only way that reversals could produce this anomaly pattern is if the rocks of the sea floor cooled through the Curie points of their magnetic minerals at different times, implying that the ages of the basaltic rocks increase with distance from the crests of the ridges. Part of the anomaly pattern from the second line of Fig. 3–13 is repeated in Fig. 3–14; with it is shown the sequence of magnetic reversals for the last few million years. The similarities in the durations of the first few epochs of a mainly normal or mainly reversed field to the widths of the corresponding belts of anomalies is striking.

In Fig. 3–14, in accordance with the Vine-Matthews hypothesis, we have

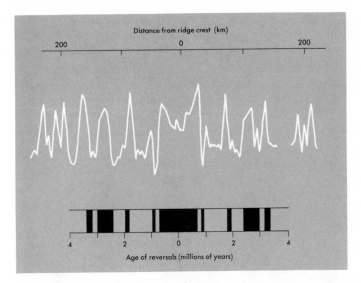

FIGURE 3-14 *Magnetic anomaly curve from the East Pacific rise (top) and the chronology of magnetic reversals (bottom). The top curve is the central part of the middle curve of Figure 3-13. At the bottom, periods of reversed polarity are shown in white, and normal polarity in black.*

plotted reversals of the magnetic field against distance in the upper part and against time in the lower. Now the physical quantity connecting distance and time is rate or velocity; something must be moving to account for the similarities we have noted. The fact that zero distance from the crest of the ridge corresponds to zero time reckoned from the present shows that it is relative motion of sea floor and ridge crest that we are observing. We cannot determine from this argument alone whether the ridge crest is migrating through the sea floor or whether the sea floor is moving off the ridge. The symmetry of the anomaly patterns, however, points strongly to the second alternative. The symmetry is automatically provided if the crust beneath the sea is forming at the ridge crests and moving away with equal velocities in both directions. This process is known as *sea-floor spreading.*

The way in which sea-floor spreading is thought to take place is illustrated schematically in Fig. 3–15. Near the ridge crest the crust and part of the mantle are continually pulling apart, creating elongated fissures that fill with molten rock. The rock that solidifies becomes magnetized in the direction of the prevailing field upon cooling. Immediately before a reversal of the magnetic field all the rocks near the ridge crest will be magnetized in the direction of the field, which for the sake of illustration we take to be as at present. Following the reversal, however, new injections of liquid will produce reversely magnetized rocks. The injections occur in a zone near the

The Earth's magnetic field

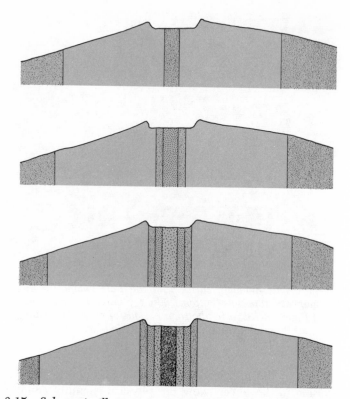

FIGURE 3-15 *Schematic illustration of the mechanism of sea-flooring spreading. Injection of new magma along cracks near the ridge crest produces a belt of rocks magnetized in the opposite sense to those injected prior to the last reversal of the Earth's field.*

ridge crest, and there is a certain degree of randomness to their distributions within this zone. Furthermore, the individual fissures extend for only a few kilometers or a few tens of kilometers along the ridge crest. They die out and are replaced by others that are distributed slightly differently with respect to the crest. The spreading motion is sporadic, occurring as bursts of activity that are localized in space and time. These small-scale irregularities in the motion account for the failure of the anomaly belts to be bounded by smooth lines that exactly parallel the ridge crests, and they also introduce the small amount of asymmetry about the ridge crest that is present in all of the magnetic profiles.

The rate of sea-floor spreading can be determined from the distances from the crest of the ridge at which magnetic reversals of known ages are found. The "half-spreading rates" of the Mid-Atlantic Ridge and East Pacific Rise are found to be about 1 cm/year and 5 cm/year respectively. (These are the rates of motion of points on the sea floor with respect to the ridge crests; two points on opposite sides of the ridges recede from each other twice as rapidly.) Spreading rates from other ridges fall in the same range. For comparison the mean rates of continental drift can be obtained from the data given in Figs. 3–7 and 3–8. They turn out to be about 1 to 2 cm/year, comparable with the rates of sea-floor spreading.

Lines of evidence other than those derived from geomagnetism support the idea of sea-floor spreading. But subsequent to the publication of the Vine-Matthews hypothesis in 1963, a rapid evolution of our ideas about global tectonics has taken place. These advances merit a chapter of their own, and it seems preferable to include them in the chapter which follows rather than here.

4

Plate tectonics

Sea-floor spreading was first suggested by H. H. Hess a few years before publication of the Vine-Matthews hypothesis. His evidence was purely geological and had nothing to do with geomagnetism. It had become clear at the time Hess wrote his paper that at the present rate of accumulation the entire thickness of sediments in the deep sea would form in less than 100 million years. (For further discussion of this point see *Oceans*, by K. K. Turekian.) Since this is only one or two per cent of geologic time, why are the sediments not thicker? Furthermore, it had been thought that somewhere at sea a complete geologic record, consisting of strata of all ages back to the time when the oceans themselves were formed, should exist. Search for such a record turned up nothing more than 100 million years old. Hess suggested that these observations could be explained if the oceanic crust formed at sea, moved across the ocean floor, and finally was swept under the continents and destroyed.

Two further lines of evidence for sea-floor spreading come from the Atlantic Ocean. The oldest lavas found on oceanic islands occur on the islands that are farthest from the Mid-Atlantic Ridge. Conversely, generally young lavas occur on islands near the Ridge. A second observation is the location of the Ridge itself. It almost exactly bisects the Atlantic, and the way it curves to follow the bulges of Africa and South America suggests that its central position is anything but accidental. Sea-floor spreading automatically provides an explanation.

Further corroboration has come from drilling in the sediments in the deep sea. Near the ridges the sedimentary cover is thin, and only the most recent sediments are present. The sediments thicken and the lowermost strata increase in age away from the ridges. The results of drilling in the Atlantic are shown in Fig. 4–1.

All of these lines of evidence join the extremely impressive magnetic evidence in support of the idea of sea-floor spreading. Further development of that idea has led to *plate tectonics*. The surface of the Earth is thought to be divided into a number of plates, each of which behaves as a more or less rigid unit. The large relative movements that we considered in Chapter 2, and others that we shall consider in this chapter, take place only at the boundaries between plates. But before considering individual plates and their relation to tectonic features, we must examine in greater detail some structures that we have already discussed briefly and also introduce a new type of structure.

FIGURE 4-1 *Age of oldest sediment, immediately above lava, as function of distance from the Mid-Atlantic Ridge. Data from Maxwell et al., 1970, Initial Reports of the Deep-Sea Drilling Project, Volume III, U.S. Government Printing Office.*

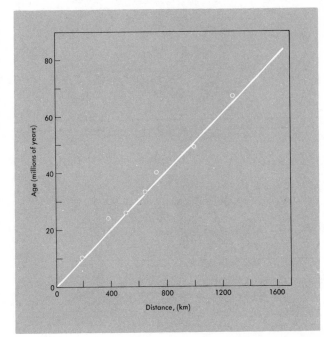

Transform Faults

Transform faults can be produced where new rock is formed at the surface by injections of igneous material. Their vital role in sea-floor spreading was recognized by J. T. Wilson. Superficially they resemble transcurrent faults, but closer examination shows that the only feature they share in common is horizontal relative movement. Otherwise they are more nearly opposites than twins. Suppose that the crest of a ridge is offset by a fault, as indicated in Fig. 4-2. If the ridge is not a source of spreading, the explanation of the offset in the topography would be motion along the fault in the direction indicated by the arrows in Fig. 4-2a. The fault is simply interpreted as transcurrent. But now consider a ridge from which spreading occurs. The motions are shown by the arrows in Fig. 4-2b. Relative motion occurs only along that part of the fault that connects the two segments of the ridge; elsewhere, the fault is simply a scar, the two sides of which move together. Furthermore, the sense of the relative motion between the ridge crests is opposite to the sense of motion in Fig. 4-2a. In the case of a trans-

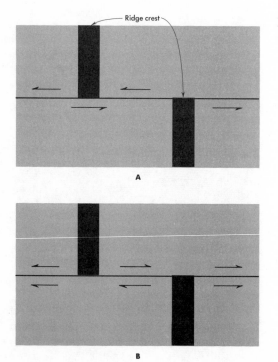

A

B

FIGURE 4-2 *(A) Transcurrent fault contrasted with (B) transform fault.*

current fault, the offset of the ridge is a consequence of the faulting. In the case of a transform fault, the faulting is a consequence of the offset of the ridge.

Transform faults enable ridge systems to change direction abruptly. The direction of spreading need not be perpendicular to the trend of the ridge, but if the plates are rigid, the rate at which new crust is created must change abruptly when the ridge changes direction. This is illustrated in Fig. 4–3; when the diagonal segment of ridge is in the same direction as the motion, no new material is added along it, and it simply becomes a transform fault. The way in which the Mid-Atlantic Ridge curves sharply around the bulges of Africa and South America with the aid of transform faults is shown in Fig. 4–4.

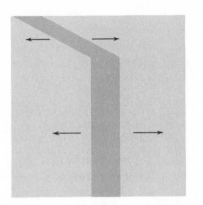

FIGURE 4-3 *The spreading rate, measured perpendicular to the ridge crest, must change when the ridge changes direction.*

It would not be possible to distinguish between transform and transcurrent faults on the sea floor were it not for the fact that motion along these faults proceeds as periodic slips, each of which causes an earthquake. The direction of the relative motion of the blocks separated by the fault can be determined from careful study of the vibrations that are produced. The motion is found to be consistent with Fig. 4–2b, not Fig. 4–2a. Furthermore, the earthquakes occur only along that part of the fault that falls between the two segments of ridge. Such a distribution of activity is precisely what we would expect from a transform fault; a transcurrent fault would not be active over so restricted a part of its length. The results just described are typical of earthquakes originating along faults offsetting ocean ridges. They show them to be transform faults, an observation that constitutes a further item of evidence supporting the idea of sea-floor spreading.

A Closer Look at Ocean Ridges

A generalized cross section of an ocean ridge is shown in Fig. 4–5. The overall width of the ridge is some 2000 km and it rises about 2 km above the surrounding abyssal plains. The topography is rugged near the crest,

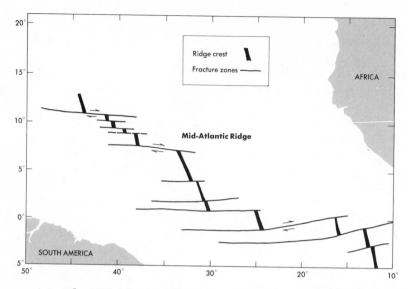

FIGURE 4-4 *The Mid-Atlantic Ridge offset by transform faults in the equatorial Atlantic. Where indicated by arrows, the faulting has been established as of the transform type by studies of earthquakes (L. R. Sykes, 1968,* The History of the Earth's Crust, *ed. R. A. Phinney, Princeton University Press).*

but it becomes smoother on the older flanks where a mantle of sediments fills the depressions. As the figure indicates, the ridge protrudes not only upward into the ocean but also downward into the mantle, as shown by the increased depth to the Moho. The Moho cannot be recognized beneath the crest of the ridge in some areas, and no sharp boundary between crust and mantle can be established in such places.

FIGURE 4-5 *Structure of the Mid-Atlantic Ridge. Densities are given for the various rock units. Vertical exaggeration 10:1. Data from Talwani, Le Pichon, and Ewing, 1965,* Jour. Geophys. Res., *v. 70, p. 341.*

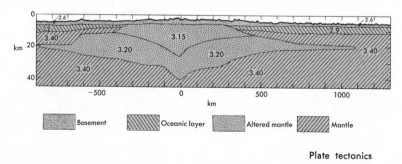

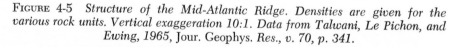

A steepsided valley is characteristically found at or near the crests of ocean ridges. Although this feature is not particularly striking when seen in a single profile such as Fig. 4–5, its persistence in profile after profile has drawn attention to it. These undersea valleys are remarkably similar in shape and dimensions to rift valleys, as shown in Fig. 4–6, which compares

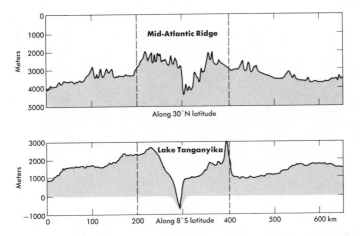

FIGURE 4-6 *The central rift of the Mid-Atlantic Ridge compared with the East African Rift at Lake Tanganyika. Vertical exaggeration 120:1. After B. C. Heezen, 1962,* Continental Drift, *ed. S. R. Runcorn, Academic Press.*

one of them to the East African Rift at Lake Tanganyika. On the basis of this similarity it has been suggested that there is a worldwide rift system lying largely beneath the sea (Fig. 4–7). Most of the injections of igneous rock accompanying sea-floor spreading are thought to occur within the central valley.

Although these undersea valleys resemble rift valleys topographically, there are other points of dissimilarity. Much of the volcanic rock in the vicinity of continental rifts has a chemical composition that is unusual elsewhere. These rocks are comparatively rich in sodium and potassium. Lavas from ocean ridges, on the other hand, possess either normal or low amounts of alkali metals. The spreading rates accompanying continental rifts must be much less than those of ocean ridges. Nothing like the rates of a few centimeters per year can be occurring in the continental rifts.

These differences raise doubts about the relation between the central valleys of oceanic ridges and continental rift valleys. The analogy between them may not be as close as sometimes has been suggested. But there can be no doubt about the importance of the central valleys. We can inspect

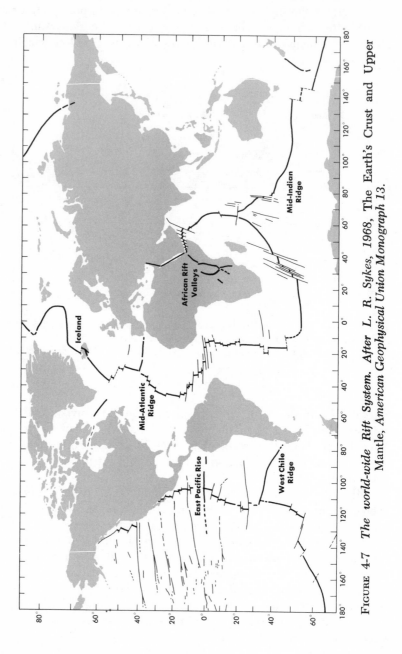

FIGURE 4-7 *The world-wide Rift System. After L. R. Sykes, 1968, The Earth's Crust and Upper Mantle, American Geophysical Union Monograph 13.*

FIGURE 4-8   *The central graben of Iceland showing a recent fault scarp and open fissures. (Photograph by Sigurdur Thorarinsson).*

one that emerges above the sea in Iceland. It is the central valley of the Mid-Atlantic Ridge, and on the island it is known as the Central Icelandic Graben. Most of Iceland's active volcanoes lie within it, and recent faults and fissures attest to the tensional forces at work (Fig. 4–8). This is certainly the most active part of Iceland, and, by analogy, of the Mid-Atlantic Ridge.

Ocean Trenches

At a spreading rate of a few centimeters per year, the ridges are creating new sea floor at an average rate of a few square kilometers per year. At this rate, the entire surface of the Earth would be produced in less than 5×10^8 years, that is, in less than one tenth of geologic time. There is no reason to suppose that spreading rates today differ drastically from the past, and so we are left with more surface produced than can be accommodated by the Earth. The excess must be destroyed somewhere; apparently the place where this occurs is in the deep ocean trenches.

Important evidence about the structure of the trenches comes from the earthquakes associated with them. Most earthquakes occur at shallow depths, but near the trenches they may occur at depths as great as 700 km. Such deep shocks are very rare in other regions. The positions of these earthquakes are systematically related to their depths, with very shallow quakes occurring almost directly beneath the topographic expressions of the trenches, and deeper ones progressively displaced away from the oceans and beneath the island arcs or continents on the opposite sides of the trenches. This relationship is shown in Fig. 4–9; it is particularly striking along the west coast of South America. (Note also in Fig. 4–9 that all earth-

Plate tectonics

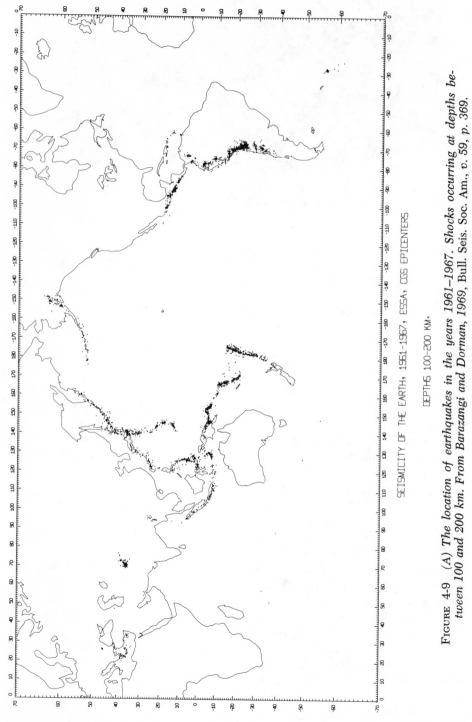

SEISMICITY OF THE EARTH, 1961-1967, ESSA, CGS EPICENTERS

DEPTHS 100-200 KM.

FIGURE 4-9 (A) *The location of earthquakes in the years 1961–1967. Shocks occurring at depths between 100 and 200 km. From Barazangi and Dorman, 1969, Bull. Seis. Soc. Am., v. 59, p. 369.*

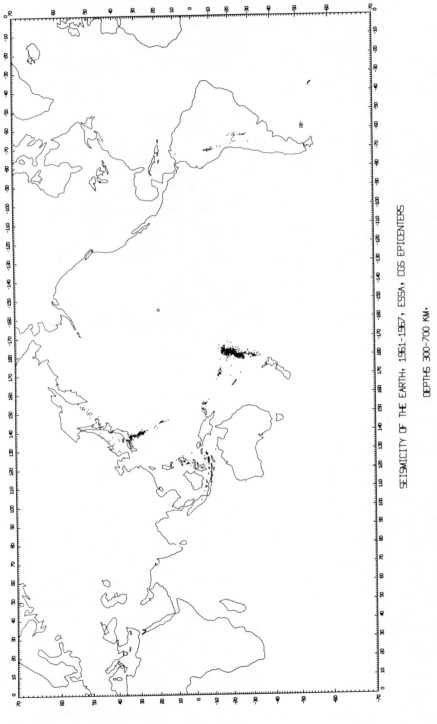

SEISMICITY OF THE EARTH, 1961-1967, ESSA, CGS EPICENTERS

DEPTHS 300-700 KM.

FIGURE 4-9 (B) The location of earthquakes in the years 1961–1967. Shocks occurring at depths between 300 and 700 km. From Barazangi and Dorman, 1969, Bull. Seis. Soc. Am., v. 59, p. 369.

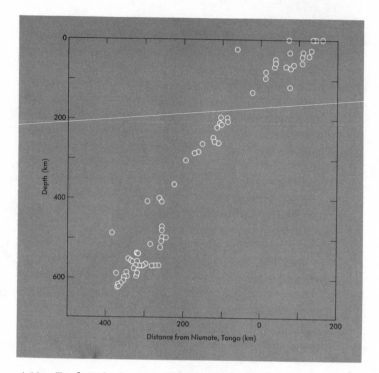

FIGURE 4-10   *Earthquake locations beneath the Tonga arc. After Isacks, Oliver, and Sykes, 1968,* Jour. Geophys. Res., *v. 73, p. 5855.*

quakes beneath the ocean ridges are shallow; none occurs at depths greater than 100 km.) Figure 4–10 is a cross section showing the locations of earthquakes beneath the Tonga Trench. They are confined to a narrow region dipping at an angle of about 45°, and such a feature is common to all trenches. The regions distinguished by the earthquakes are called *Benioff zones,* after Hugo Benioff, who first noticed them. The vibrations produced by the earthquakes show that the Benioff zones are sinking relative to their surroundings. Most of the world's active volcanoes occur in the island arcs and continental margins above the Benioff zones. The idea that crust is destroyed in the Benioff zones can be fitted with other features of plate tectonics to make a coherent picture. A slab of sea floor, including the entire oceanic crust and some of the underlying mantle, is drawn downwards along the Benioff zone and eventually digested by the deeper mantle (Fig. 4–11). The drag of this descending material produces the topographic depression that we observe as the trench. We are faced with an apparent paradox, however, for it appears that the descending tongue of cool, near-surface material gives rise to volcanoes. Part of the solution to this problem is to be found in the chemistry of the descending tongue. It may contain

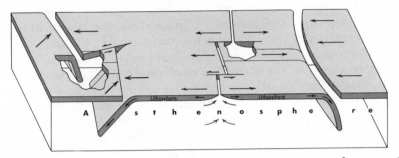

FIGURE 4-11 *The dynamics of plate motion, showing creation of plates at a ridge and destruction in trenches. After Isacks, Oliver, and Sykes, 1968, Jour. Geophys. Res., v. 73, p. 5855.*

entrapped water that contaminates the surrounding mantle and lowers its melting point, producing molten rock. In addition, friction between the sinking material and its surroundings produces heat that raises the temperature of both the Benioff zone and the adjacent mantle, thereby causing further melting. An important part of the process of digesting the Benioff zone entails fusion and the removal of liquid, some of which reaches the surface to form volcanoes.

Mountain Building

It is commonly observed that the deformed rocks comprising mountain belts originated as unusually thick sequences of sediment deposited under marine conditions. As early as 1859, James Hall originated the concept of the *geosyncline,* the name given the trough that permits the accumulation of great thicknesses of sediment by subsidence. Further work showed that the history of a geosyncline is complicated and that tectonic activity is prevalent throughout. Portions of the trough may rise above the sea as islands for a time, suffer erosion and shed sediments into the surrounding depths, and then subside to become sites of deposition once again. Volcanic activity may also contribute material, but not necessarily to all parts of the geosyncline. Eventually the rocks of the geosyncline are compressed into mountains and uplifted above the sea, and the geosyncline is destroyed as a site of sedimentation.

The role of the geosyncline as a precursor of mountains and the complexity and diversity of the orogenic process can be explained within the framework of plate tectonics. The ideas expressed here are new, and some might consider them tentative. But observational support is increasing rapidly. Although many details remain to be worked out, the scheme to be outlined offers real insight into the orogenic process and its causes.

A central role is played by the descending tongue of cold material that consumes crust and by the margins of continents. Continental margins may broadly be classified into Atlantic and Pacific types. The former is relatively simple; earthquakes and volcanoes are absent, and a trough of sediment is formed by erosion of the adjacent continent. Pacific margins are more diverse. There is the Cordilleran type, of which the Andes are an excellent example, and the type of margin characterized by an offshore island arc and a marginal sea, such as the Japanese islands and the Sea of Japan. The interaction between a descending tongue and continental margins and island arcs cause the orogenic process.

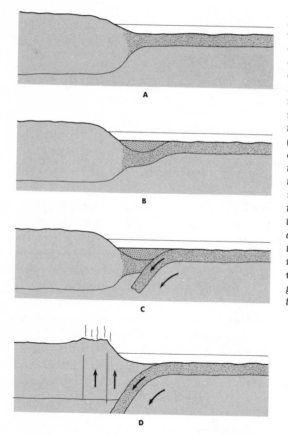

FIGURE 4-12 *The conversion of an Atlantic-type coastline into a Cordilleran type. (A) Original continental margin. (B) Development of a thick trough filled with sediment, partly in response to the sedimentary load. (C) Disruption of the oceanic crust and initiation of a descending tongue. (D) Partial melting of the descending tongue. Rising liquids heat the overlying crust and deformation accompanies the heating. Igneous and metamorphic activity convert the continental margin into a mountain system that eventually becomes new continental crust.*

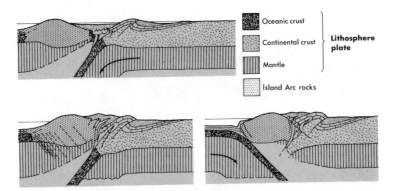

Oceanic crust
Continental crust
Mantle
Island Arc rocks

Lithosphere plate

FIGURE 4-13 *Continent-island arc collision, followed by change in the direction of the descending tongue. After Dewey and Bird, 1970,* Jour. Geophys. Res., v. 75, *p. 2625.*

Formation of a new descending tongue can transform a continental margin of Atlantic type into a Cordilleran mountain system. A way in which this can be accomplished is shown schematically in Fig. 4–12. The oceanic crust beneath the sediment-filled trough becomes unstable, fractures, and initiates a descending tongue. A possible reason for this instability is discussed in a later section. The drag of the descending tongue compresses the sediments, producing folds and faults, and liquids generated during the digestion of the descending tongue rise, carrying with them enough heat to produce partial melting of the overlying continental crust and sediments. At the conclusion of the orogeny, the new mountain belt is left as an extension of the original continent.

A second way in which a descending tongue may evolve into a mountain belt is through what may be termed collisions, although the processes are so slow that no real impact occurs. Collision occurs because anything attached to an oceanic plate that is being consumed, such as a continent, must eventually arrive at the trench above the descending tongue. In this manner the continent can be pulled against an island arc or another continent. As we shall see, the continental crust is composed of light rocks, and it is too buoyant to be pulled down by the tongue. The tongue therefore disappears; it may be replaced by a new one dipping in the opposite direction in the case of a continent-island arc collision, as illustrated schematically in Fig. 4–13. Figure 4–14 shows a continent-continent collision. There are a number of ways in which a descending tongue can interact with a continental margin to produce the diverse types of orogenic belts that we observe.

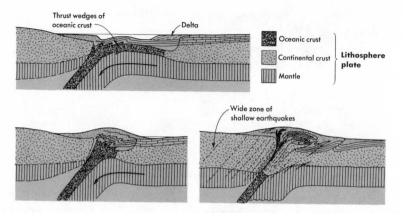

FIGURE 4-14 *Continent-continent collision. After Dewey and Bird, 1970,* Jour. Geophys. Res., *v. 75, p. 2625.*

Motions of Present-Day Plates

The structural elements that bound plates at the surface are of three types. The first is the system of ridges, with associated transform faults, where plates are created. The second consists of the trenches where plates are destroyed, and we include with them young fold mountains, such as the Alps, that are interpreted as products of the interaction between a descending tongue and a continental margin. The third structural element is a connecting network of transcurrent faults that neither creates nor destroys plates, but enables them to slide past each other. The pattern of major plates at the Earth's surface is shown in Fig. 4-15. Details of the pattern remain to be worked out, and in some complex regions such as the eastern Mediterranean, where there appears to be a number of small plates, the interpretation is still incomplete.

Plates come in a range of sizes. The large ones, such as the American plate or the Pacific plate, are of continental dimensions. Others, such as the Caribbean plate and the Arabian plates, are subcontinental in size, and they range on downwards to much smaller plates. As an illustration of plate tectonics, we consider the motion of the large Pacific plate, shown in Fig. 4-16. The main source of this plate is the East Pacific rise and its continuation across the southern ocean to a point southwest of New Zealand. North of the Gulf of California the boundary is the San Andreas system of transform faults. Complications set in between northern California and the United States-Canadian border because a small plate intervenes between the larger American and Pacific plates. The boundary of the Pacific plate

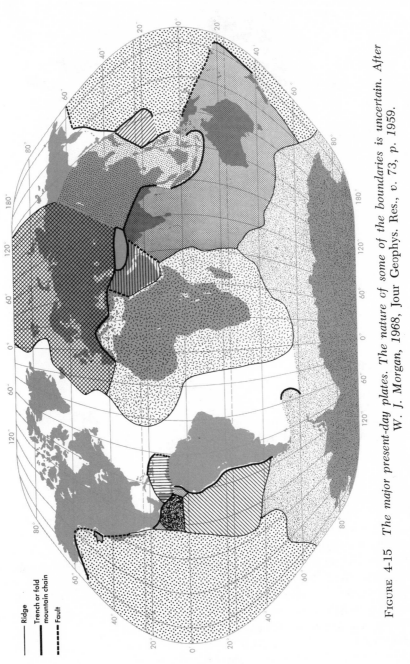

FIGURE 4-15   *The major present-day plates. The nature of some of the boundaries is uncertain. After W. J. Morgan, 1968, Jour Geophys. Res., v. 73, p. 1959.*

Ridge
Trench or fold
mountain chain
Fault

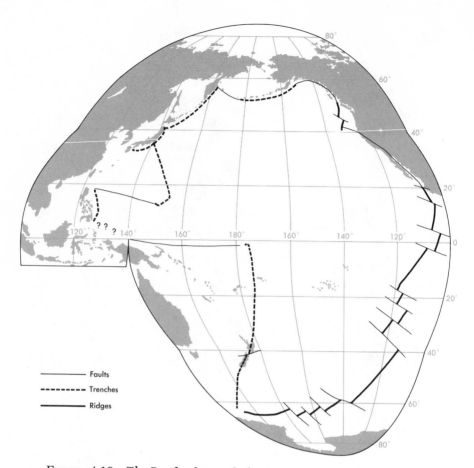

FIGURE 4-16   *The Pacific plate with the tectonic elements that bound it.*

turns abruptly westward in the Gulf of Alaska and changes from a trans-current fault to the Aleutian trench. This is the easternmost extension of the line along which crust is destroyed. The trench continues westward along the Aleutian chain, but because of the curvature of the arc the trench gradu-ally gives way to another transcurrent fault, which extends to the Kamchatka peninsula. Here the boundary again becomes a trench, which persists to central Japan, where another complication is encountered. The Bonin-Marianas trench strikes southward from Japan, outlining another small plate between the Asian and Pacific plates. From here to the well-defined trench extending from Tonga to New Zealand the boundary is poorly known. A transcurrent fault seems to be required to accommodate the northwestward motion of the Pacific plate, but northward motion of the Indian-Australian plate (Fig. 4–15) requires a trench (the Java trench) along this boundary as well. From Tonga the trench-type boundary proceeds southwestwards to rejoin the ridge in the southern ocean.

This interpretation of the tectonics of the Pacific plate is generally consistent with all known geological and geophysical data, and it makes a coherent picture. Where ambiguity or difficulty is present, it is due either to insufficient information or to situations that other observations (such as magnetic measurements) show to be very complicated. It is encouraging to find that the hypothesis of rigid plates can apparently account for the tectonics of the Pacific, which comprises about a quarter of the surface of the Earth.

The Pacific plate also carries evidence of a different direction of motion relative to North America in the past. A number of prominent fracture zones have been recognized in the eastern Pacific (Fig. 4–17), and their orientations relative to each other are consistent with their being vestigial scars of transform faults originating on a ridge system to the east. Their

FIGURE 4-17 *The major fracture zones of the northeast Pacific. After H. W. Menard, 1967,* Science, *v. 155, p. 72.*

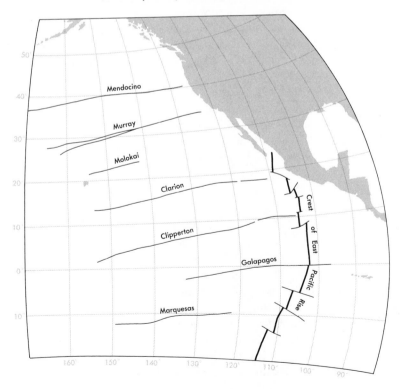

Plate tectonics

orientations are inconsistent with the present direction of spreading from the East Pacific rise, nor is there a ridge to the east of the four most northerly fracture zones at present. Apparently what we are observing is evidence of previous movement of the Pacific plate away from the American plate in a direction south of west towards New Guinea and the Philippines. At present, the relative motion is northwestwards towards Japan and the Aleutians. The old ridge off the Pacific coast of North America must have disappeared beneath the continent, which is possible only if there was a trench as well as a ridge near the coast. There is evidence in the geology of the Pacific Coast ranges that they were once associated with a trench, and much of the geological activity of the North American Cordillera, which we noted in Chapter 1, may be produced by the continent overriding a once-active ridge.

The southern margin of the Eurasian plate and part of the margin of the East Asian-Indonesian plate are bounded by the high fold mountains of the Alpine-Himalayan chains. A great seaway, known as the Tethys trough, formerly existed where the mountains are now. The eastern end of the Tethys received geosynclinal sediments from Cambrian through Eocene times, but in the western, or European, end, geosynclinal conditions did not commence until the Permian period. Whether a trench existed in the Tethys trough during the time of sedimentation is not known, but the large movements of the African and Indo-Australian plates relative to the Eurasian plate, that are suggested by paleomagnetism (Figs. 3–7 and 3–8), would be easier to understand if a trench had been present at least part of the time in the Mesozoic and Tertiary periods.

Plate tectonics provide a solution not only to the problem of the large compressions we observe in fold mountains but also to the problem of the termination of transcurrent faults of large displacement. The large relative motions are just what we should expect if large-scale creation and consumption of the Earth's crust is taking place at the inferred rate.

The Driving Mechanism

Up to now we have deliberately avoided discussion of the thicknesses of the moving plates. We have no direct evidence bearing on this question, but rough inferences can be drawn from results that will be discussed in later chapters. The conclusion is that the plates are 50–100 km thick in the oceans and somewhat thicker, perhaps up to 200 km, in the continents. These figures are rough but of adequate precision for present purposes. The main point is that the plates are thin in comparison with their horizontal dimensions, which are of the order of 10,000 km in the case of the large plates. The lower boundary must be very well lubricated, sufficiently to reduce the coefficient of friction there sensibly to zero, if such thin bodies are to be moved without disruption.

The term *lithosphere* is used to describe the material making up the plates. Originally it referred to the entire rocky part of the Earth, but the recent tendency is to restrict it to the rigid outermost layers. Beneath the lithosphere lies the *asthenosphere*, which is the weak layer that provides the lubrication. Although this term is much older than plate tectonics, its usage here is consistent with its original definition. The lithosphere should not be confused with the Earth's crust. The former is distinguished by its strength under conditions of very slow deformation, whereas the latter is distinguished by its short-term elastic responses.

Our ideas about mechanisms that drive the plates about are still vague, but gravity plays a central role. The lithosphere at sea may be a colossal gravity slide moving off of the elevated ridges and into the abyssal depths. In addition, a series of mineralogical transformations, the tectonic importance of which has been emphasized by A. E. Ringwood, may play a crucial role. The lavas that emerge at the ridge crests to form the oceanic crust have the chemical compositions of rocks known as *basalts*. Crystallizing a liquid of basaltic composition under near-surface conditions produces a rock consisting of nearly equal proportions of the minerals feldspar and pyroxene, with a density of about 2.9 gm/cm^3. (See *Earth Materials*, by W. G. Ernst, for a description of these and other rocks and minerals mentioned in this section.) If a basaltic rock is subjected to higher pressures owing to increasing depth of burial and weight of overburden, the minerals of which it is composed change in response, and the name by which it is called changes as well. More important, the density of the rock increases progressively with pressure. The sequence of rocks formed by subjecting basalts to increasingly high pressure is shown in Table 4–1. The overall composition of the rock is the same throughout the table, but the compositions of the individual minerals, as well as their proportions, change with the increasing pressure.

The overall effect of the transformations given in Table 4–1 is the conversion of a basalt to an eclogite, with the granulites appearing as intermediate stages. As we shall see, basalt is less dense than any likely mantle rock, whereas eclogite may well be denser than the rocks representative of the uppermost mantle. Thus, an oceanic crust of basalt would simply "float" on the underlying mantle, whereas a chemically identical crust of eclogite would be mechanically unstable and tend to sink because of its higher density.

Ringwood's suggestion is that whenever the oceanic crust becomes buried deeply, it may be converted first to granulite and then to eclogite. It would then sink into the mantle and initiate a descending tongue and an associated

Table 4–1

Some Rocks Formed by Metamorphism of Basalt

Name	Approximate Mineralogy	Density
Amphibolite	75% hornblende 25% feldspar	3.0
Pyroxene granulite	55% pyroxene 45% feldspar	3.0
Garnet granulite	55% pyroxene 25% feldspar 20% garnet	3.2
Eclogite	60% garnet 40% Na-bearing pyroxene	3.6

trench. Such deep burial might occur beneath the sedimentary trough of an Atlantic-type continental margin and provide the instability required to produce a Cordilleran-type mountain system. It might also occur beneath a volcanic archipelago, where the necessary depth of burial is provided by a thick pile of volcanic rocks. In the latter case, the descending tongue can be initiated within an ocean, far from a continental margin.

5

Seismology

Seismology is the study of the vibrations generated by earthquakes or by large explosions. The commonest sources of vibrations are earthquakes, which are associated by most people with the death and destruction that can be caused by a large one occurring near a populated region. But scientists are far more interested in the much smaller vibrations that are generated several times daily. These vibrations are too small to feel, but they can be detected by sensitive instruments. As we have seen, earthquakes occur where masses of rock are moving slowly. Pressures build up until they are released with explosive suddenness, and part of the energy that had been stored in the deformed rock is carried away by the resulting vibrations. The phenomenon is rather like the whipping motion imparted to a rope that is stretched until it snaps.

The point or the restricted region in which the seismic energy is released is called the *focus* or *focal region*. The point on the surface directly above the focus is called the *epicenter*. The position in the Earth at which an earthquake occurs is specified by giving the latitude and longitude of the epicenter and the focal depth.

Seismic Waves

The vibrations generated by a seismic source travel outwards through the Earth. From the outset it must be realized that the individual particles of rock do not move any great distance. Instead, they oscillate about their undisturbed positions. The motion and the associated energy, however, are passed from particle to particle and travel long distances. These vibrations, or waves, are transmitted through the Earth with finite velocity. Consequently, for a time after the earthquake has occurred, a surface, known as the *wave front*, separates material that has been set into vibration from undisturbed material that the wave has not yet reached. The wave front moves outwards from the focus with the velocity of propagation of the waves.

Seismic waves are elastic disturbances that deform the solid elastically during their passage. In an isotropic material there are two types of waves which travel with different velocities and deform the solid in different ways. A material is *isotropic* if its properties, in this case the velocity with which a wave travels, are the same in all directions. Cast iron is a good example of an isotropic material, and wood is a good example of material that is not isotropic. The first variety of elastic wave, which is called a *P-wave* because it causes the first or *primary* disturbance, deforms the material in the manner indicated in the first row of Fig. 5–1. A cube of material is alternately lengthened and shortened in the direction of propagation of the wave. Its other dimensions are unchanged, and the angles at the corners remain right angles. P-waves are also termed *compressional* waves because the volume of the cube is changed; it is alternately compressed and expanded.

The second variety of elastic wave is called an *S-wave* because it produces the *secondary* disturbance. A cube is deformed as indicated in the second row of Fig. 5–1. The front and back faces are distorted into parallelograms having the same areas as the original square face. The other faces are unchanged in size or shape. S-waves do not change the volume of the cube; they simply change its shape. They are also termed *distortional* or *shear* waves for this reason.

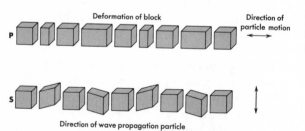

Deformation of block

Direction of particle motion ←→

P

Direction of wave propagation particle

S ↕

FIGURE 5-1 *Deformations produced by the passage of P-waves and S-waves.*

The direction of particle motion caused by each type of wave, which can readily be deduced from the deformations of the cubes, is shown on the right side of Fig. 5–1. The motion is back and forth along the direction of propagation of a P-wave; S-waves produce motion at right angles to the propagation direction. The *push-pull* motion of P-waves and the *shake* motion of S-waves provide a way of remembering the motion associated with each variety. The terms *longitudinal* and *transverse* are also applied to the two types of waves to describe the nature of the particle motion.

P-waves resemble sound waves, and S-waves resemble light. Like light, they can be polarized. An S-wave can be considered to be made up of two components: one, called an *SH-wave*, vibrating in a *horizontal* direction; and the other, called an *SV-wave*, vibrating *vertically*.

The P- and S-waves pass through the interior of the Earth; they are therefore termed *body waves*. There is another class of elastic waves, called *surface waves*, that only disturb the material close to the surface. Surface waves, however, are really a special case of a more general type of vibration that affects the whole Earth. The term *free oscillations* is used because the Earth as a whole is set into vibration like a gigantic bell, free of the influence of external forces or constraints.

Free Oscillations of the Earth

Before we discuss the free oscillations of the Earth, it is helpful to consider a simpler example. If we stretch a string between two fixed supports, we can cause it to vibrate by displacing it from its equilibrium position and suddenly setting it free. We could choose to make the shape of the tsring quite complicated before releasing it; if we did so, we would find that its shape was complicated at later times as well. Nevertheless, there exists a system of simply shaped displacements that can be superimposed on each other to reproduce the complicated shape of the string exactly, both at the instant of release and at subsequent times. These simpler displacements, which are called *normal modes*, are shown in Fig. 5–2.

The time required for the string to swing through a full cycle is called the *period* of the vibration. The periods of the normal modes shown in Fig. 5–2 are the times required for the string to go from the positions shown in black to the white positions and back again. Each normal mode of vibration occurs with a single period that is characteristic of that mode and is not shared by any other. The normal modes are unique in this respect; vibrations of more complicated shapes are characterized by as many discrete periods as there are normal modes making up the complex vibration. Furthermore, these periods are the ones of the normal modes. By analysis of a gen-

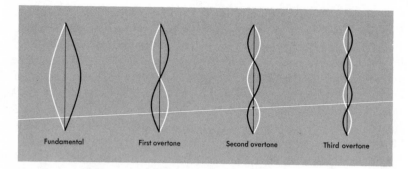

FIGURE 5-2 *Normal modes of vibration of a string. The string moves from the position shown in black to the position shown in white, and back again.*

eral, possibly complicated, motion, the periods of the normal modes can be determined. These in turn give information about the properties of the string and the tension in it.

The normal modes have in common the feature that an integral number of half-wavelengths exactly equals the length of the string, thus assuring that the ends of the string remain fixed. Figure 5–2 shows that, except for the fundamental mode, there are other points on the string that do not move. They are called *nodes*, which should not be confused with *modes*. The fundamental, or lowest, mode has no nodes; the first overtone, or second normal mode, has one node; and so on. The higher the mode, the larger is the number of nodes and the shorter the period.

The basic concepts we have just discussed in connection with the vibrating string are also applicable to the more complicated case of the vibrating

FIGURE 5-3 *Radial modes of vibration of a sphere. Internal nodal surfaces are indicated by dashed circles.*

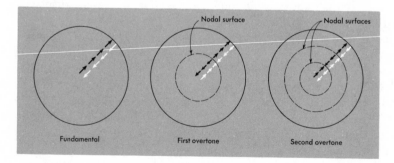

sphere. A complex general motion can be resolved into its normal modes, each of which has its own characteristic period. Measurement of these periods gives information about certain properties of the sphere. Nodes, which were points on the string, are surfaces on which there is no displacement in the case of the sphere. Furthermore, there are two distinct classes of oscillations of a sphere, which for reasons that will emerge are called *spheroidal* and *torsional* oscillations.

The simplest free oscillations of a sphere are the *radial* modes, which represent a special class of spheroidal oscillations. The sphere retains its spherical shape and simply expands and contracts. (Fig. 5–3). There are no nodal surfaces in the fundamental mode; in the overtones the nodes are spheres concentric with the outer surface. They are confined to the interior of the sphere and do not intersect its surface.

More general types of spheroidal oscillations are shown in Fig. 5–4. The volume of the sphere is changed by all oscillations of the spheroidal type,

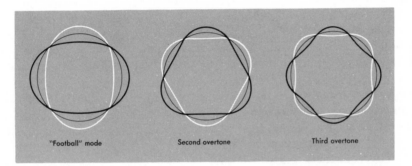

"Football" mode   Second overtone   Third overtone

FIGURE 5-4 *Spheroidal vibrations of a sphere. Points at which the black and white lines intersect represent nodal lines in the surface of the sphere. The fundamental mode of spheroidal vibration is represented by the radial mode of Figure 5-3.*

and there is a radial component to the motion, as is shown by the displacements of the outer surfaces in Figs. 5–3 and 5–4. Overtones of the spheroidal modes of Fig. 5–4 may have internal nodal surfaces like those shown in Fig. 5–3; they may also have nodal surfaces that intersect the surface of the sphere to produce nodal lines. Modes with a large number of nodal lines at the surface are said to be *high-order* modes.

Torsional oscillations involve twisting motions of the sphere, whence their name. The motion has no component in the radial direction; it is purely horizontal. Consequently, the spherical shape is preserved, and the volume

FIGURE 5-5 *Torsional vibrations of a sphere. Nodal lines in the surface are indicated.*

of the sphere is unchanged. The directions of motion at the surface are shown in Fig. 5–5. In the lowest mode indicated in the figure, the northern hemisphere twists one way and the southern the other; in the next mode, the polar regions twist the opposite way from the equatorial region. Once again there are nodal lines produced by the intersection of nodal surfaces with the sphere's surface, and they increase in number as the order of the mode gets higher. It is also possible to have internal nodal surfaces with torsional oscillations. In the simplest case, the sphere twists one way at depths shallower than the node and the other way at greater depths (Fig. 5–6).

High-order modes of free oscillation have relatively short periods, and the motion is concentrated close to the surface. These modes may also be regarded as surface waves. It is to be emphasized that the surface waves are completely equivalent to high-mode free oscillations, and, conversely, the lower modes of free oscillation are equivalent to long-period surface waves with wavelengths that may exceed the radius of the Earth. Surface waves were recognized before the lower modes of free oscillation because their higher amplitude makes them easier to observe and because instruments capable of detecting the very long-period oscillations have been developed only comparatively recently.

FIGURE 5-6 *Torsional vibration showing an internal nodal surface.*

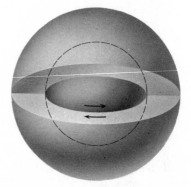

Two types of surface waves, corresponding to the two classes of free oscillation, exist; they are named after the men who interpreted them theoretically. *Rayleigh* waves, named for Lord Rayleigh, are spheroidal oscillations, and *Love waves,* named for A. E. H. Love, are torsional oscillations. In Rayleigh waves the motion of particles is a mixture of P- and SV-wave type; the particles move in ellipses, as indicated in Fig. 5–7. In Love waves the motion is like SH-waves.

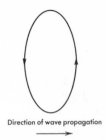

Direction of wave propagation

FIGURE 5-7 *Particle motion during passage of a Rayleigh wave.*

Travel Times of Body Waves

The primary data that are obtained from the study of body waves are the lengths of time required for P- and S-waves to travel from the focus to points of observation at various distances away. In seismology distance is conventionally measured as the number of degrees of arc separating the epicenter from the point of observation (Fig. 5–8). Once the travel times are determined as functions of distance, the velocities of P- and S-waves as functions of depth in the Earth can be derived. A major goal of observational seismology is to determine these velocities as accurately as possible.

At first glance, it might appear to be impossible to derive travel times from earthquakes because we do not know *a priori* where or when the shock occurred. But this is not true; we can take a rough, preliminary set of travel times and by a process of successive approximations refine them to

FIGURE 5-8 *Distance* Δ *as measured in seismology.*

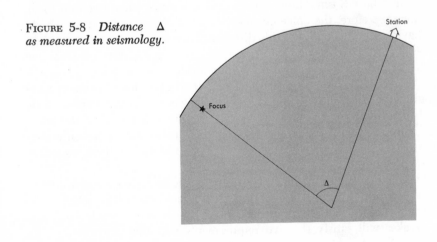

give accurate results. Data from explosions, particularly nuclear blasts, greatly assist in this process, since the uncertainty in place and time of the shock is negligible. Waves from a large nuclear explosion can be recorded all over the world and provide valuable data.

Artificial explosions help in working out travel times but they are not essential. Suppose that we start with no knowledge of the times. There are places in the world, for example, California, Japan, or southern Europe, that are densely populated, have frequent earthquakes, and where a network of closely spaced seismograph stations has been built. In such regions earthquakes are felt fairly commonly, and we can take the point where the most intense effects are reported as a preliminary estimate of the location of the epicenter. Distances to local observatories can be read from a map, and extrapolation of the observed times of arrival of P- and S-waves to zero distance gives the time of the earthquake. We now have all the information required to construct tables or graphs giving the travel times of P- and S-waves. We should expect the observations to show considerable scatter about our predicted curve, and we should also expect the departures to be systematically related to the azimuth between the epicenter and the observatory. For example, all of the stations to the north of the epicenter might show a greater time interval than predicted, whereas those to the south might show a smaller interval. Such a pattern would mean that our trial epicenter is too far to the north (see Fig. 5–9). We then try an epicenter further to the south and eventually find by trial and error the location that causes the observed P-wave and S-wave times to fall as nearly as possible on a smooth curve versus distance. A new set of travel times can now be calculated using the improved location of the epicenter.

Once local travel times are available for a given region, we have a way of locating the epicenters of all earthquakes, whether or not they are felt by people. The P-S time interval tells us the distance from a station to the epicenter; therefore, the epicenter must lie on a circle, centered on the station, having a radius equal to this distance. A circle constructed in a similar way about a second station will generally intersect the first circle at two points. A third circle about a third station located the epicenter unambiguously (Fig. 5–10). Note that in the figure the three circles define a small triangle rather than intersecting at a point. This is due to small errors in the travel times and errors in measurement of the times of arrival of the waves at the three stations. This epicenter, found by using data from only three stations, can be refined with data from more stations in the manner described previously.

Once the epicenter and time of an earthquake have been established by a local network of seismograph stations, observations of its P- and S-waves all over the world will determine the travel times. Only a small number of earthquakes will satisfy the two requirements that they occur where the

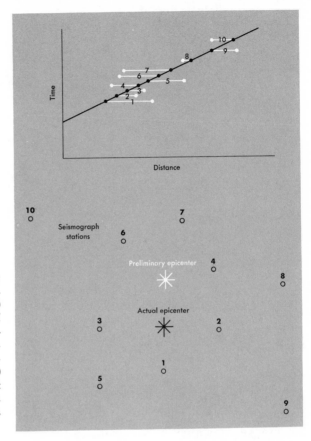

FIGURE 5-9 *Refinement of epicentral location. Initial trial epicenter (white) led to the badly scattered white points on the travel–time plot. By trial and error a location of the epicenter was found (black) which produced the black points on the plot through which a smooth curve can be drawn.*

epicenter can be accurately located and that they be well-recorded all over the world. But a few particularly suitable shocks will provide a worldwide set of preliminary times.

We define the difference between the time at which a wave is actually observed at a seismograph station and the time predicted from our preliminary set of travel times as the *residual* at that station. To observe systematic trends, we examine the residuals for a large number of earthquakes at a large number of stations. We might find, for example, that they are largely positive between angular distances of 60° and 80° and negative between 80° and 100°. Such behavior indicates that our preliminary travel times are too small in the first interval of distance and too large in the second. Therefore, we must revise our times to give, as nearly as possible, zero mean residuals at all distances. We next redetermine the epicenters and

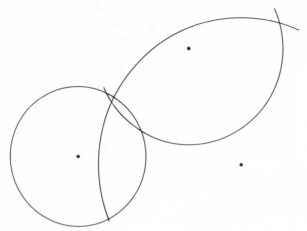

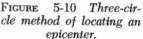

FIGURE 5-10 *Three-circle method of locating an epicenter.*

times of the earthquakes in order to secure the best possible agreement with the new times, recompute the residuals, and again look for systematic trends. The process is repeated as many times as necessary; fortunately, only about two repetitions are required, for a large amount of labor is involved.

In our discussion so far, we have tacitly assumed that the earthquake occurred at shallow depths. Such is not always the case, and we require criteria for recognizing deep-focus shocks and determining their focal depths. One criterion for recognition is that deep-focus earthquakes generate unusually small surface waves, although they produce body waves that closely resemble those from shallow shocks of comparable magnitude. Extra waves are often observed from deep-focus earthquakes; they represent waves that move upwards from the focus, are reflected at the Earth's surface near the epicenter, and then travel to the point of observation (Fig. 5–11). The time intervals between these waves and the normal P- and S-

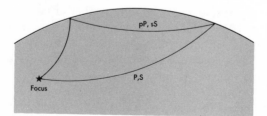

FIGURE 5-11 *Paths of extra waves (termed pP and sS) that are observed in the case of deep-focus earthquakes.*

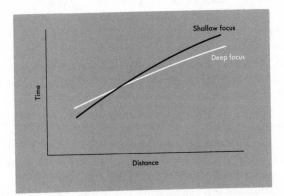

FIGURE 5-12   *Travel times for shallow and deep-focus earthquakes. Waves from the deep-focus shock arrive late near the epicenter and early at great distances.*

waves that have traveled directly to the observer can provide a good measure of the focal depth. These extra waves are not observed from shallow shocks because they arrive too close to normal P- and S-waves to be distinguished.

The travel times for deep-focus earthquakes differ systematically from the times for a surface focus. Near the epicenter the waves from a deep shock arrive late because they travel a longer path. But at intermediate and large epicentral distances the focus is closer than the epicenter to the observatory because of the curvature of the Earth (Fig. 5–12). Hence, waves from deep shocks arrive early, and the discrepancy with times from shallow sources increases with distance. Such systematic discrepancies provide an additional means of recognizing deep shocks. Once the times from shallow earthquakes have been determined, it is an easy matter to work out the corrections required for various focal depths. Observed residuals against the predicted times for a surface focus can then be used to determine the depth of a deep-focus earthquake.

Behavior of Body Waves at an Interface

We have already introduced the concept of the wave front as the surface separating material that has been disturbed by the passage of the wave from undisturbed material. Lines perpendicular to the wave front, which are in the direction of advance of the wave front and also in the direction in which energy is carried by the wave, are called *rays*. When waves are generated in material of constant velocities of P- and S-waves, the wave fronts are spheres, and the rays are straight lines radiating outwards from the focus. But when the velocities of the waves are not the same everywhere, the rays become curved lines; this phenomenon is known as *refraction*.

The laws that govern the refraction of seismic waves provide a means of deducing the seismic velocities in the Earth from observed travel times. We cannot go into general aspects of refraction theory here, but it is helpful to consider the paths of rays in a special case and to derive the resulting travel times. Suppose that we have a spherical shell with a constant velocity V_1 lying on a central spherical region with constant velocity V_2. We need not specify whether we are considering P-wave or S-wave velocities. Within each region of constant velocity the ray paths are straight lines, but they change directions at the interface. The refraction is governed by *Snell's Law* which, like many concepts of seismic ray theory, was originally stated as a principle of optics. A ray incident upon the interface at an angle θ_1 with the normal (Fig. 5-13) gives rise to a reflected ray at the same angle and a refracted ray at an angle θ_2 that is related to θ_1 by the relation $\sin \theta_1 / V_1 = \sin \theta_2 / V_2$, according to Snell's Law. Actually, in the case of an elastic wave the situation is more complicated; an incident ray gives rise to four rays: reflected P- and S- and refracted P- and S-. But unless the change in velocity at the interface is large, most of the energy from an incident P-wave goes into P-waves. Similarly, S-waves mainly produce S-waves. In the interior of the Earth, the only important place where P-waves are converted to S-waves, and vice versa, is at the outer boundary of the

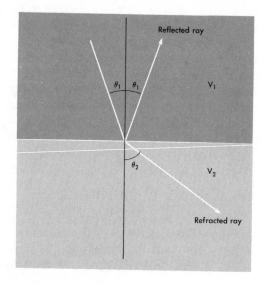

FIGURE 5-13 *Reflection and Snell's law of refraction at an interface between media in which the velocities are different.*

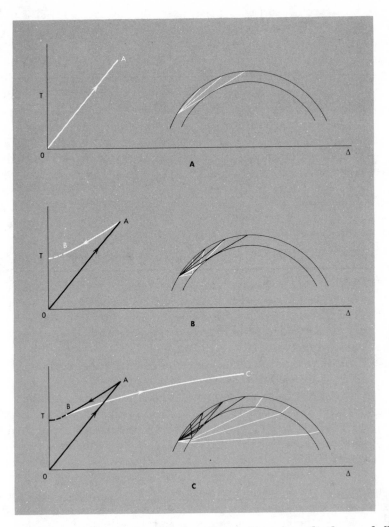

FIGURE 5-14 *Refraction and reflection at an interface where the deeper shell has the higher velocity. As the angle of incidence decreases, the three branches of the curve are traced out in parts (A), (B), and (C).*

core. Nowhere else are such conversions significant, and we shall neglect them for the remainder of this section.

The effect of an interface on travel times depends on whether the higher velocity is in the shallow shell or the deep central region. We first consider the case of $V_1 < V_2$ (Fig. 5–14). The direct path through the upper layer terminates with the ray that grazes the interface at the base of the shell (Fig. 5–15a). Rays penetrating slightly deeper strike the interface; their angle of incidence (θ_1 in Fig. 5–13) is near $90°$, and hence $\sin \theta_1$ is nearly 1. Since $V_2 > V_1$, these are rays for which Snell's Law gives $\sin \theta_2 > 1$, an impossibility. Such rays cannot be refracted. Instead, they are totally reflected, and their point of emergence moves back towards the focus as the angle of incidence decreases. This behavior is opposite to that of the direct rays, and for this reason the part of the travel time curve corresponding to total reflection (shown in Fig. 5–14b) is termed a *receding branch*. As indicated by the dotted line in the figure, reflections can be observed back to zero distance. They are strongest in the interval between A and B however, since none of the energy of the incident ray goes into a refracted ray in that range of distances. Eventually, an angle of incidence for which Snell's Law gives $\sin \theta_2 = 1$ is reached, and for smaller angles refraction through the central region takes place. The points of emergence of the rays again increase in distance with decreasing angle of incidence, giving rise to the third branch of the travel-time curve (Fig. 5–14c).

The discontinuous rise in velocity at the interface produces triplication of the travel-time curve; that is, at distances between the points A and B of Fig. 5–14 three different rays that have traveled different paths arrive at different times. If the second and third arriving waves can be observed, they confirm the existence of a discontinuous change in velocity, but the ground has already been set in motion by the first wave, making observation of later waves difficult.

The travel-time curve has an entirely different appearance when the lower velocity is beneath the interface, that is, when $V_1 > V_2$. A branch resulting from direct transmission of rays through the upper shell occurs as before (branch OA of Fig. 5–15), and again it terminates with the ray that grazes the interface. Deeper rays are refracted downwards, away from the surface, and emerge at the distance B in Fig. 5–15. With further decrease in the angle of incidence, the rays trace out the receding branch BC, emerge at a minimum distance at the cusp C, and follow the normal branch CD. No rays emerge at distances between A and C; this interval is called a *shadow zone*. Duplication of the travel-time curve occurs between C and B. Total reflection cannot occur in this case, and the actual reflected rays may be weak.

If the layer of low velocity is thin and underlain by higher-velocity material, waves refracted through the lower layer may emerge in the shadow zone. There may be no ray that reaches its greatest depth in the low-velocity layer; that is, all rays that enter it penetrate it completely. Under these circumstances it can be difficult even to detect a low-velocity layer, to say nothing of the difficulties in determining its thickness and velocity. The ambiguities that can be encountered are illustrated in Fig. 5–16, in which two very different velocity distributions, one with and one without a low-velocity layer are shown. The quantity $T\text{-}12\Delta$ is plotted as the ordinate in the figure because it permits the vertical scale to be greatly expanded over the scales

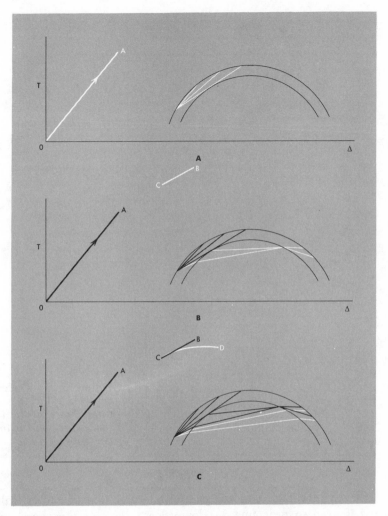

FIGURE 5-15 *Refraction at an interface where the deeper shell has the lower velocity. As the angle of incidence decreases, the three branches of the curve are traced out in parts (A), (B), and (C).*

in Figs. 5–14 and 5–15. The black and white curves would be indistinguishable on the scale of the latter figures. The low-velocity zone is twice as thick as the overlying layer of higher velocity, and the drop in velocity is a full 0.5 km/sec as the low-velocity zone is entered in the structure shown as black lines in Fig. 5–16. The simpler structure represented by the white lines yields travel times that never depart by more than 10 seconds from the times in black. Some of the largest differences occur amongst the late ar-

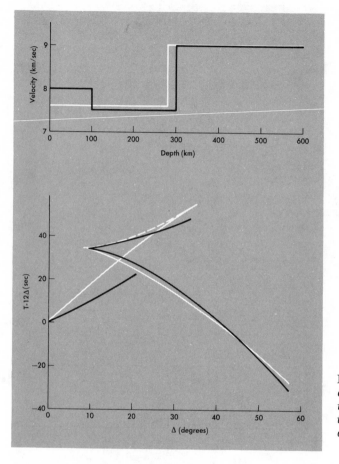

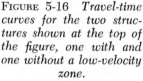

FIGURE 5-16 *Travel-time curves for the two structures shown at the top of the figure, one with and one without a low-velocity zone.*

rivals. Clearly, accurate data and careful interpretation of late arrivals are required if even so pronounced a low-velocity zone as the one shown in Fig. 5–16 is to be unambiguously located.

The influence of discontinuities in velocity on the ray paths and travel times provides a way of inferring qualitatively what to expect from continuously varying velocities, since a continuous distribution can be approximated by a large number of small, closely spaced, discontinuities. If the velocity always increases with depth, the ray paths curve upwards, as shown in Fig. 5–17. We note that the BC branch in Fig. 5–14 is flatter than the OA branch, which is a consequence of the higher velocity below the interface. We can, therefore, infer that with continuously increasing velocity the travel-time curve is concave towards the depth axis, as Fig. 5–17 shows. The points A and B of Fig. 5–14 approach each other as the discontinuity in velocity becomes small; as a consequence, there is no triplication of the travel

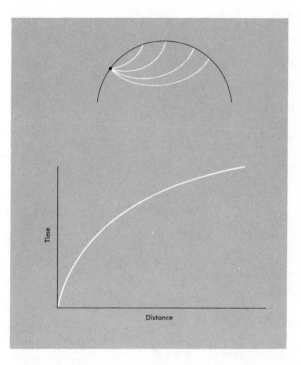

FIGURE 5-17 *Ray paths and travel times when the velocity increases continuously with depth.*

times. If a thin low-velocity zone is present, it alters the shapes of the ray paths considerably, and the travel-time curve is nearly straight for an interval. The effect on the times need not be very great however. A thick zone of low-velocity, on the other hand, would produce a shadow zone.

Travel Times and the Velocity Distribution in the Earth

The times of P- and S-waves are shown in Fig. 5–18. They are based on observations of thousands of earthquakes that occurred in the seismically active regions and that were recorded all over the world. Fitting the best possible curves to these observations, without regard for the particular path traveled by the waves, leads to an Earth model in which the velocities are assumed to depend on depth only and not on latitude and longitude at a given depth. If this assumption were not made, Japanese earthquakes recorded in Europe, for example, would have to be treated separately from South American earthquakes recorded in North America, and the curves would be based on much less data. Combining all the observations represents a first step leading to an average Earth model. Determining such regional contrasts as may exist is a refinement that has not yet been completed.

The travel-time curves in Fig. 5–18 are concave towards the distance-

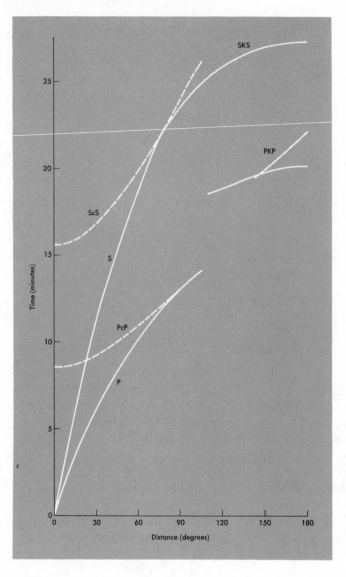

FIGURE 5-18 *Travel times in the Earth. Branches labeled PcP and ScS represent reflections from the core-mantle boundary, and those labeled PKP and SKS represent waves that penetrate into the core.*

axis, indicating a general increase in velocity with depth. Careful inspection of the figure shows that the strongest curvature in the P- and S- branches is near 20°, implying a region where the velocity increases particularly rapidly. But the most striking feature of Fig. 5–18 is the termination of the P- and S- branches at about 105°, followed by a shadow zone in P-waves. This behavior constitutes the principal evidence for the existence of the Earth's core. If the times of P-waves observed at less than 100° are extrapolated to 180°, the predicted time falls short by a few minutes of the observed time at that distance, an observation that led R. D. Oldham to first infer the existence of a core with lower velocity.

Further confirmation of the core's existence comes from observations of reflections, the times of which are labeled PcP and ScS in Fig. 5–18. They show that the margin of the core is marked by an abrupt change in velocity rather than a gradual change over an interval of depths. In the latter case, reflections strong enough to observe would not be produced. Waves refracted through the core are labeled with a K in Fig. 5–18. The branch PKP corresponds to P-waves above the core, while the branch SKS corresponds to S-waves. The branches S, ScS, and SKS combine to give the form of triplication characteristic of a discontinuity with the higher velocity below. P and PKP, on the other hand, show the shadow characteristic of a lower velocity below the discontinuity. The PKP and SKS branches prove to represent waves that traveled with the same velocity in the core; there is no evidence for separate P- and S-waves with different velocities. Liquids and gasses cannot transmit S-waves, and their absence is the principal evidence that at least the outer part of the core is liquid. Only P-waves can travel through it, and SKS gives an example of a wave converted from S-wave to P-wave and back again.

Waves that have traveled through the core were observed and identified by Beno Gutenberg, and his work confirmed Oldham's suggestion. But further complications appeared. The two branches of PKP beyond the shadow zone intersect at 143°, and strong motion is commonly recorded near that distance. However, weaker motion can be traced as far back as about 110°; it would not be found at distances shorter than the intersection of the two branches if there is only a single discontinuous decrease in velocity (compare Fig. 5–15). I. Lehmann accounted for these observations by assuming that a second discontinuity, with the higher velocity beneath, occurred inside the core.

The travel-time curve for PKP resulting from Lehmann's hypothesis is shown schematically in Fig. 5–19. The branches ABC correspond to the branches BCD in Fig. 5–15, and the branches $BCDEF$ correspond to $OABC$ in Fig. 5–14. Only the branches AB and DEF, shown in black in Fig. 5–19, are well-observed. Events that do not fall on any of the branches are rather commonly recorded; the most mysterious of these are the "precursors," which are weak motions occurring earlier than the branch DEF. There is also uncertainty as to the distance of the point D. It is shown at 110° in Fig. 5–18, but some observers believe that it is located at greater distance, perhaps as much as 125°. Motions at shorter distances are then interpreted as reflections, prolonging the DEF branch in the manner of the dashed curve in Fig. 5–14. Thus, there is still uncertainty in the times of waves passing near the Earth's center; hence, it is difficult to establish velocities at great depths.

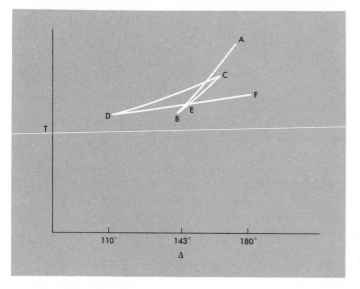

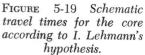

FIGURE 5-19 *Schematic travel times for the core according to I. Lehmann's hypothesis.*

Duplication of travel times at short distances was first noticed by A. Mohorovičić in a study of waves from a Croatian earthquake, and the discontinuity causing it bears his name. The times resemble the branches *OA* and *BC* of Fig. 5–14; the reflected branch is seldom, if ever, observed.

Two extensive studies of travel times, one by Sir Harold Jeffreys and K. E. Bullen, and the other by Gutenberg, led to two classic velocity distributions. In both, the velocities were assumed to vary as smoothly as possible with depth. Discontinuities were only introduced where duplication or triplication of the times, or reflections, confirmed their presence. The results are shown in Fig. 5–20, and it is clear that the two studies are in broad general agreement. The results are virtually indistinguishable at depths between 600 and 2900 km, and both agree that the velocity of P-waves drops from 13.6 to about 8 km/sec when the core boundary is crossed.

However, in addition to these broad points of agreement, there are important differences between the two studies. Jeffreys gives 2900 and 5120 km for the depths to the core and the deeper discontinuity discovered by Lehmann, whereas Gutenberg gives 2920 and 5080 km. The Jeffreys velocity rises more rapidly in the outer part of the core but reverses its trend and drops as the lower discontinuity is approached. Gutenberg's curve rises in the outer part of the core, changes slope and rises steeply at a radius of 1290 km, and drops slightly near the center of the Earth. The two solutions differ by 0.2 km/sec at the center, and their maximum difference in the core is about 1 km/sec at the lower discontinuity of Jeffreys.

In view of the uncertainties in the travel times, which were discussed earlier, it is not surprising to find differences of opinion about the velocities

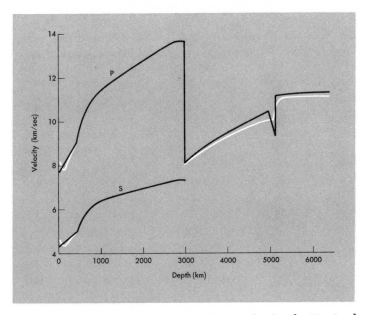

FIGURE 5-20   *The Jeffreys (black) and Gutenberg (white) velocities in the Earth. Where only one curve is shown the two sets of data agree too closely to distinguish on the scale of the figure.*

inside the core. Later workers have proposed even more complicated variations of velocity, with many discontinuities that give the core a layered structure. Other than their comparative complexity, these later proposals have little in common, however, and the velocity deep in the core has yet to be satisfactorily determined.

In addition to the differences in the core, the Jeffreys and Gutenberg velocities disagree in the upper part of the mantle in a most important fashion. Jeffreys shows the velocities increasing with depth, whereas Gutenberg found them to decrease with depth near the Moho, pass through minima at depths near 100 km, and then rise. The two sets of velocities approach each other at greater depths and become indistinguishable in the mantle below about 600 km. Minima in the velocities are features with far-reaching consequences, as we shall see, and it is **very** important to know whether or not they exist in the upper part of the mantle. But, as we have emphasized, thin zones of low velocity are very difficult to detect because their effect on travel times is small. No clear reason to prefer one set of velocities over the other could be found in the times, and for a number of years the issue remained in doubt. A decisive test, which clearly favored Gutenberg's results, was obtained by studying the free oscillations of the Earth.

The velocities in the Earth's crust vary both with depth and from locality to locality. P-waves travel between 5–7.4 km/sec, and the velocity of S-waves is 3–4 km/sec. Almost all of the good determinations of velocity have come from studies of waves generated by explosions, rather than earthquakes, and the velocity of P-waves is better known than of S-waves because the former arrive earlier and are easier to observe. Typical crustal sections for stable continental regions and for the deep ocean are shown in Fig. 5–21. Beneath the Moho the velocities at the top of the mantle are usually between 7.9–8.2 km/sec for P-waves and between 4.7–4.8 km/sec for S-waves.

For two reasons the first determinations of the velocity distributions were derived from study of body waves rather than free oscillations. Long-period seismographs capable of detecting the free oscillations represent a technological advance that was made comparatively recently. Equally important, the interpretation of observations of the free oscillations presents a formidable computational problem that could not be tackled prior to the development of modern high-speed digital computers. Once the necessary tools became available however, it was possible to supplement the data from travel times with a new body of information, namely the periods of the various modes of oscillation.

Like the derivation of velocities from travel times, the interpretation of the free oscillations is done by some trial-and-error method. A commonly used procedure is to assume that the Earth consists of nested spherical shells, each having constant properties, separated by discontinuities. The properties in each shell are varied until agreement with observed periods is secured over the entire range of the observations. Different modes of oscil-

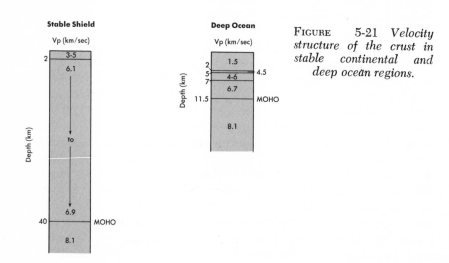

FIGURE 5-21 *Velocity structure of the crust in stable continental and deep ocean regions.*

lation affect the Earth to varying depths; consequently, changing the properties of a particular layer produces shifts in period that vary over the spectrum of observed periods. It is possible to fix the properties of a few hundred individual layers by using the full range of observations.

The periods of torsional oscillations depend on the velocities of S-waves and the densities in each layer. Periods of spheroidal oscillations depend on P-wave velocities as well. The sensitivity of the periods to the properties of a layer depend in a complicated fashion on the depth of the layer and on the particular mode under consideration. Only the lowest modes are sensitive to properties at great depths. In general, the periods are most sensitive to the velocities of S-waves, which implies that these must be accurately known before P-wave velocities and densities can be derived. The free oscillations supplement body waves in the sense that the travel times are best-known for P-waves.

The results of analysing the free oscillations definitely favor Gutenberg's velocities over those of Jeffreys. A minimum in S-wave velocity occurs in the upper mantle all over the world, but we cannot be as sure about P-waves. Not only do the latter have less effect on the periods of free oscillations, but also a minimum in P-wave velocities seems to be a local feature, present in some regions but absent in others. All of these results depend on an assumed distribution of density, but the sensitivity of the periods to density is too low to leave the conclusion about the minimum on the S-wave velocity in doubt.

It is possible in principle to solve for both of the velocities and for the density, given the periods of free oscillations and the travel times of P- and S-waves. This possibility is being realized. Progress is slow, however, because the insensitivity of the observed quantities to density requires that very accurate data be obtained. The accumulation of such data takes time, and the process is still continuing. A modern velocity distribution in the mantle, based on much more complete data than was available to Jeffreys and Gutenberg is shown in Fig. 5–22. The conspicuous new features are the nearly discontinuous increases in velocities at depths of about 200, 400, and 650 km, which replace the sharp but uniform rise in velocity shown by the older curves in this range of depths. Minima in both P-wave and S-wave velocities occur between 100–150 km. The velocities deeper in the mantle do not change smoothly with depth. Instead, they show irregularities in their rates of increase with depth, particularly between 800–1000 km and 2600–2800 km. The velocity of S-waves actually decreases nearly discontinuously between 2750–2775 km, but the other irregularities are too minor to be discerned clearly in the figure. It should be emphasized again that the velocities shown in Fig. 5–22 represent worldwide averages that do not necessarily represent any particular region.

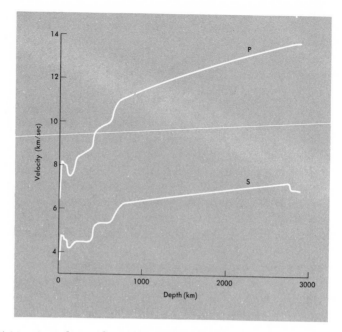

FIGURE 5-22 *A modern velocity distribution in the mantle. Data from Anderson and Kovach, 1969, Bull. Seis. Soc. Am., v. 59, p. 1667.*

Seismic Attenuation

When the Earth is set into vibration by an earthquake, the motion does not persist indefinitely. Instead, the vibrational energy is gradually converted into heat, and the Earth comes to rest. The motion is then said to be *attenuated,* and the process causing the attenuation is called *anelasticity* or *internal friction.*

Study of attenuation requires that the amplitudes of seismic waves be measured; that is, we must determine the distance that the ground is displaced by the wave. But the displacement is affected by local geological conditions and thickness of soil cover, and it varies from place to place in a complicated fashion. In practice, most successful studies of attenuation have been made under conditions permitting a wave to be observed several times at the same station. Several cycles of free oscillations can be observed after strong earthquakes, and the decrease in amplitude, for example, can be measured. A station near the epicenter of a deep-focus earthquake can record direct waves, waves reflected from the core, waves reflected first at the outer surface and then by the core, and so on. Comparison of the amplitudes of these events permits the attenuation below the focal depth to be compared with the attenuation at shallower depths.

The interpretation of studies of attenuation is still somewhat uncertain, so we shall restrict ourselves to broad generalities. The attenuation is low in the crust, rises to a maximum in the upper part of the mantle, and falls to extremely low values in the mantle below 600 km. The maximum attenuation seems to occur in the low-velocity zone. The significance of these results has been aptly expressed by D. L. Anderson, who says that attenuation is "not unrelated" to the strength of the material in which it is observed. By this he means that in a qualitative sense high attenuation implies low strength and vice versa (lead and other weak materials would make poor bells). The attenuation data for the Earth imply that the low-velocity zone is also a zone of low strength. It is here that we should find the asthenosphere, which permits large-scale horizontal movements. Over the asthenosphere is the stronger lithosphere and under the asthenosphere lies what may be the strongest material in the Earth, the mantle material deeper than about 600 km.

The constitution of the Earth

from seismic evidence

Previously, we had considered the Earth divided into crust, mantle, and core. The seismic velocities described in the last chapter form the basis for the finer subdivisions of the Earth given in Fig. 6–1. The three-fold division of the mantle and two-fold division of the core can be inferred from the Jeffreys and Gutenberg velocities; the more recent results shown in Fig. 5–22 confirm the subdivision of the mantle, and the depths of the boundaries are based on these data. The reason for the name *transition zone* in the mantle will emerge later in this chapter. We have seen that there is reason to suppose the core is more complex than indicated in Fig. 6–1, but it is still premature to consider anything but the two-fold division.

The velocities of P- and S-waves in materials of known constitution can be measured in the laboratory. Such data can be used to interpret the constitutions of the crust and upper mantle by direct comparison. But the enormous pressures encountered at greater depths alter the velocities so profoundly that such direct comparison is impossible. It is more useful to consider other properties that can be calcu-

Subdivision	Depth to Interface, Km
Crust	
	11–50
Upper mantle	
	350
Transition zone	
	750
Lower mantle	
	2890
Outer core	
	5000
Inner core	

FIGURE 6-1   *Major subdivisions of the Earth.*

lated from the velocities, because we have a better understanding on the basis of both theory and experiment of how they are affected by pressure. We can extrapolate these properties to lower pressures with considerable confidence and then make comparison with experimental data. But before considering the velocities and related properties, we shall consider the pressures in the Earth.

Pressures in the Earth

In a liquid Earth, the decrease in pressure with radius would be given by the relation

$$dp/dr = - g(r)\rho(r),$$

where P is pressure, r is the distance from the center, $g(r)$ is the acceleration of gravity, and $\rho(r)$ is the density. This equation expresses the change in the weight of overburden, per unit area, which equals the change in pressure in a liquid. In the solid parts of the Earth, however, some of the weight of overburden may be supported by strength, and the actual pressure may be different from that given by the simple equation above. But the strength of the Earth is limited; in the crust it can hardly exceed 1000 bars, and in the upper mantle the data from seismic attenuation indicates considerably weaker material. The lower mantle may be far stronger, but this makes little difference because we are really interested in the strength in relation to the pressure calculated from the equation above. Thus, although the absolute error in calculating pressure from the equation may be large, amounting to tens of thousands of bars, it is still a small fraction of the pressure existing throughout much of the Earth's interior. Near the top of the mantle the pressure is about 10,000 bars, and the uncertainty due to strength is no more than 1000 bars or 10 per cent. A pressure of 100,000 bars is reached at a depth of only about 300 km and the uncertainty is no more than a few per cent. The relative deviations in pressure decrease because of the much higher pressures at greater depths.

Figure 6–2 shows the pressures calculated from a particular density distribution. Gravitational theory enables g to be calculated from the densities, and the relation between g and ρ is such as to make the pressures relatively insensitive to the density distribution actually chosen. The pressures are accurate to within a few per cent except near the surface where the effect of strength is important. The highest sustained pressure that can currently be achieved in the laboratory is no more than 200,000 bars, equivalent to a depth of less than 600 km or less than one-tenth the distance to the center.

FIGURE 6-2 *Pressure in the Earth. A megabar is roughly one million atmospheres.*

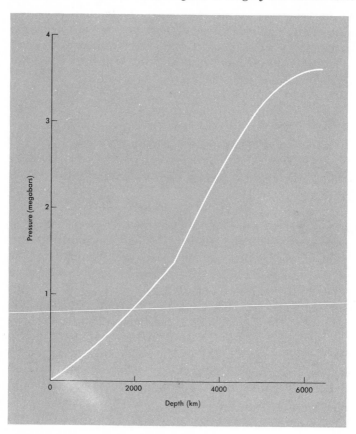

Shock waves can produce much higher pressures, which only last for about one-millionth of a second however. They are produced when a sample of material is struck a very heavy blow, either by attaching it to a block of high explosive and then detonating the charge, or by firing a projectile into it from a high-velocity gun. Both pressure and density can be measured during the brief duration of the shock. Pressures of the order of that at the center of the Earth can be produced in this way. The results of shock-wave experiments will be discussed later in this chapter.

Increase of Density by Self-Compression of a Homogeneous Layer

The pressure increases with increasing depth in the Earth because of the increase in the weight of overburden. As a result, the material is compressed more and more, and its density rises. If we know the seismic velocities, we can calculate the increase in density, provided we are dealing with a homogeneous layer. By *homogeneous* we mean that the material is identical everywhere in the layer in its *uncompressed state*. In the layer, as we observe it, the density increases downwards because of the rise in pressure and compression.

The calculation of the change in density is shown in Fig. 6–3. The final result derived there is known as the *Williamson-Adams equation*.

Inhomogeneity of the Earth's Mantle

On the assumption that the mantle is homogeneous, we can find its density by solving the Williamson-Adams equation. We have all of the information needed to do this, since φ (see Fig. 6–3) can be calculated from the known velocities, and g can be determined from gravitational theory and the densities that are determined as the calculation proceeds. Routine numerical integration gives the densities, but it must start from an assumed density at a single depth. This problem was first attacked by Williamson and Adams, but they were unable to obtain much useful information because good velocity distributions were not available at the time. Bullen reopened the matter by using the vastly improved velocities that Jeffreys had derived. In addition, he recognized that two quantities constrained his results. They are the Earth's mass and its moment of inertia, both of which are known from measurements that will be discussed in Chapter 7. Both quantities can be calculated from a density distribution, and only those distributions yielding the correct results can be accepted.

$$\text{P-wave velocity: } V_P = \left(\frac{K + 4/3G}{\rho}\right)^{1/2}$$

$$\text{S-wave velocity: } V_S = \left(\frac{G}{\rho}\right)^{1/2}$$

where K = bulk modulus
G = shear modulus
ρ = density

So
$$V_P^2 - \frac{4}{3}V_S^2 = \frac{K}{\rho} = \frac{dP}{d\rho}.$$

Let
$$\frac{K}{\rho} = \phi.$$

If
$$dP = -g\rho dr,$$

$$-\frac{g\rho dr}{d\rho} = \phi$$

or
$$\frac{d\rho}{dr} = -\frac{g\rho}{\phi},$$

the Williamson-Adams equation.

FIGURE 6-3   *Derivation of the Williamson-Adams equation.*

It is worthwhile digressing briefly to discuss the moment of inertia of the Earth. The moment of inertia, I, of any spherical body can be written in the form $I = zMR^2$, where M is the mass of the sphere, R is its radius, and z is a constant. In the case of a spherical shell, which has all of its mass concentrated at the surface, $z = \frac{2}{3}$. If all of the mass of the sphere is concentrated at the center, $z = 0$, and if the density of the sphere is the same everywhere, $z = \frac{2}{5}$. Thus, the moment of inertia gives a measure of the degree to which the mass is concentrated towards the center, through the factor z. If z is greater than $\frac{2}{5}$, we may infer that the central part of the sphere is less dense on the average than the outer part, whereas we can infer the opposite if z is less than $\frac{2}{5}$. In the Earth $z = 0.3306$; hence, the average density must become higher towards the center.

Bullen assumed that the density was 3.32 gm/cm³ at the top of the mantle, a value appropriate to the mineral olivine which he took to be the major

The constitution of the Earth from seismic evidence

constituent of the upper mantle. He calculated the densities down to the core and determined the mass and moment of inertia of the mantle from his results. He further assumed that the density was constant in the crust and found that its mass and moment of inertia was so small that their effects on his subsequent calculations were also small. Subtracting the masses and moments of inertia of crust and mantle from the corresponding quantities for the Earth as a whole leaves the mass and moment of inertia of the core. Bullen found that the factor z defined in the last paragraph was equal to 0.57, which is closer to the value for a spherical shell than for a uniform sphere. The implication is that the density of the core decreases markedly with depth.

Bullen's result was wholly unacceptable, as he realized. The seismic evidence is that the outer core is liquid. But a heavy liquid lying on a liquid less dense is unstable in a gravitational field. Either the liquids overturn and the lighter one emerges on top, like oil floating on water, or they mix and the density increases with depth. The inner core is too small to offer a means of avoiding the difficulty. If the core is hollow, with zero density in the inner core, $z = 0.42$ for the whole core if the outer core has uniform density. This is still much less than Bullen's result, and it is inconceivable that the core is in fact hollow. For the whole core, z almost certainly lies in the range 0.385 to 0.390, and we must reexamine the assumptions upon which the calculation of mantle densities are based in order to find a way out of the dilemma posed by Bullen's results.

One of Bullen's assumptions that is open to question is his choice of 3.32 gm/cm³ at the top of the mantle. This figure must be increased to nearly 3.7 gm/cm³ if z is to be brought down to 0.4 in the core. Few rocks have so high a density, and the ones that have the observed velocities in addition are rarer still. If they are predominant in the uppermost mantle, they should be found at the surface far more frequently than in fact they are. It seems clear that the cause of Bullen's discrepancy must lie elsewhere.

A second and subtler cause of difficulty might lie in the fact that a very definite assumption about the behavior of temperatures in the Earth is well-concealed in the Williamson-Adams equation. A more careful and sophisticated derivation of the equation is required to bring this out, but we need not concern ourselves further. The correction for the departure of the temperatures in the Earth from the implicitly assumed conditions has the wrong sign. It increases the value of z found for the core, rather than decreasing it, and therefore makes the discrepancy worse.

By a process of elimination Bullen arrived at the conclusion that the mantle is inhomogeneous. Subsequently, Francis Birch demonstrated that the increase in velocity on passing through the transition zone is far too great to be attributed to self-compression of homogeneous material. He suggested that the familiar minerals found at the surface undergo a series of transformations in the transition zone, through which they achieve close atomic packing in response to the high pressures. Such changes in mineralogy are termed *phase transformations*. It is possible for them to occur without change in overall chemical composition, but there is the further possibility that the chemical composition of the Earth changes progressively on passing through the transition zone. Before we can proceed further, we must consider further evidence about the densities in the Earth.

Density Distribution in the Earth

Bullen's results demonstrated that the density distribution in the mantle cannot be found from the Williamson-Adams equation alone. Birch's findings showed that the equation is inapplicable in the transition zone. He also showed that it cannot be used in the upper mantle, either because the temperatures depart from those assumed in deriving the equation, or because the region is inhomogeneous, or for both reasons. The periods of the free oscillations currently afford the best means of calculating the density. But because the periods are relatively insensitive to density, as discussed in the last chapter, it is difficult to obtain high precision, and only recently have the data achieved useful reliability. The requirements that the density distribution must lead to the correct mass and moment of inertia prove to restrict severely the acceptable range of density. Nevertheless, the densities are not as well-known as the velocities, and they are shown in Fig. 6–4 as a range of possibilities rather than as a definite curve. Despite the uncertainty remaining, much useful information can be deduced from the data in the figure.

Constitution of the Crust and Upper Mantle

In these parts of the Earth the pressure is within the range of laboratory experimentation, and the velocities can be compared with directly measured values. Experience shows that the velocities in a wide variety of common rocks fall in the range observed in the crust. Many studies of the velocities in the crust indicate that they increase slightly with depth. This is partly the effect of increasing pressure, but it can also imply a progressive change in chemical composition with depth. Elements such as sodium, potassium, and silicon may be concentrated towards the surface, whereas magnesium and iron are more abundant at depth. This point has important implications about temperatures in the crust, and we shall return to it in Chapter 8.

However, when we try to find rocks with velocities matching the ones ob-

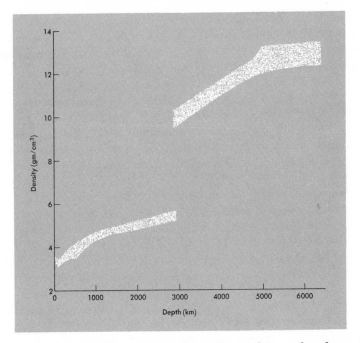

FIGURE 6-4 *Densities in the Earth. A range of possibilities rather than a definite curve is shown.*

served in the upper mantle, we find that our choice is much more limited. Rocks are aggregates of minerals, and the velocities in them are averages of the velocities in their mineral constituents. A list of velocities in the common mineral groups is given in Table 6–1. The reader is referred to *Earth Ma-*

Table 6–1

Density and Velocities of Common Minerals and Mineral Groups

Mineral	Density, gm/cm³	Velocity, km/sec		φ, km²/sec²
		P-wave	S-wave	
Quartz	2.65	6.3	3.9	19.4
Feldspar	2.56–2.76	6.76–7.10	3.73–3.81	27.2–31.0
Amphibole	3.14	7.36	4.1	31.8
Pyroxene	3.27	7.95	4.63	34.6
Olivine (Mg-rich)	3.33	8.35	4.79	39.1
Olivine (Fe-rich)	3.75	7.36	3.90	33.9
Garnet	4.18	8.52	4.77	40.0

The constitution of the Earth from seismic evidence

terials, by W. G. Ernst, for details about the mineralogy and chemistry of these minerals. Comparison of Table 6–1 with Fig. 5–22 shows that only olivine and garnet have velocities higher than those observed in the upper mantle. The feldspars, which are the commonest minerals near the Earth's surface, have velocities strikingly lower than shown by the mantle. The velocities indicate that this group of minerals is a minor constituent of the upper mantle, if present at all. Such a deduction is consistent with the observation that feldspars are unstable at the pressures of the upper mantle, decomposing to pyroxenes and quartz. The rarity or absence of feldspar is the chief mineralogical characteristic distinguishing the mantle from the crust.

The seismic velocities do not lead to a unique interpretation of the constitution of the upper mantle however. A wide variety of mixtures of olivine, pyroxene, and garnet, with or without minor amounts of feldspar, would satisfy the seismic results. Two models of the constitution of the upper mantle are of particular interest because they seem the most plausible in the light of studies of rocks showing evidence of having originated at great depths. One model holds that the mantle is peridotite, a rock consisting dominantly of olivine, with lesser amounts of pyroxene and possibly small amounts of garnet. On the second model, the upper mantle is eclogite, a rock consisting of roughly equal quantities of pyroxene and garnet. The peridotite model implies that the mantle is chemically distinct from the crust, whereas this is not necessarily true on the eclogite model. Eclogites may, as we have seen, have chemical compositions similar to basalts, which are the commonest types of lava. The mineralogy of eclogite, pyroxene and garnet, differs from the pyroxene-feldspar mineralogy of basalts because eclogite crystallized at high pressure. The lower crust may be basaltic in composition, and if the eclogite model of the upper mantle is correct, the Moho could represent a change in mineralogy without a change in overall chemical composition.

A decisive test of these two models would be provided by an accurate determination of the density of the upper mantle, for eclogite is distinctly denser than peridotite. Present data on density are not precise enough to enable us to decide on this basis.

Further mineralogical zoning is indicated by the behavior of the velocities around the low-velocity zone. According to Fig. 6–4, a minimum in density accompanies the minima in velocity, indicating either the presence of an inherently light layer differing in composition from its surroundings or further changes in mineralogy. An interesting possibility is that partial melting takes place at these depths. This could occur at a reasonable temperature if the mantle contains a small amount of water. In the presence of water, liquid may form hundreds of degrees below the dry melting temperature, but if the amount of water is small, the amount of liquid will also be small until the dry melting temperature is approached. A partly molten layer would have almost no strength at all as long as it is deformed slowly, and it could provide the lubrication required for sea-floor spreading. We would also expect the high seismic attenuation that is observed in this part of the Earth.

Constitution of the Transition Zone and Lower Mantle

As a means of studying these transformations occurring in the transition zone, we consider their product, the lower mantle. Pressures here are 400,-000 bars and higher, which is beyond the range of our experience. It is necessary to extrapolate the properties of the lower mantle to low pressures in order to compare them with the same properties measured in the laboratory. All observable properties, and their variation with pressure, are not equally well-understood, and we should choose to extrapolate those we understand best if we are to have maximum confidence in our results. The properties best suited for extrapolation are the density and the parameter φ defined in Fig. 6–3. The results of extrapolation to room temperature and pressure are $\rho = 4.25 \pm 0.1$ gm/cm^3 and $\varphi = 60 \pm 2$ (km/sec)^2. Most of the uncertainty arises because our knowledge of temperature in the lower mantle is limited. Values of ρ and φ for common minerals are given in Table 6–1, and the same properties of simple oxides are given in Table 6–2. None of the common silicates that make up almost the whole of the crust and upper mantle has properties corresponding to those of the lower mantle, but it is possible to devise a mixture of simple oxides with the requisite values of ρ and φ.

The atoms making up the simple oxides are closely packed together, giv-

Table 6–2

Densities and Elastic Ratios of Simple Oxides at Room Temperature and Pressure

Mineral	Density, gm/cm³	φ, km²/sec²
$MgAl_2O_4$	3.62	55.8
MgO	3.58	45.3
CaO	3.29	32.2
Al_2O_3	3.99	63.2
Fe_2O_3	5.27	39.2
TiO_2	4.25	51.8
SiO_2 (stishovite)	4.29	85.3
BeO	3.00	73.4

ing them high densities and low compressibilities. This combination of properties leads to high values of φ. Note also that stishovite, the high-pressure form of SiO_2, has higher values of ρ and φ than are appropriate to the lower mantle. In its structure the silicon atoms are in six-fold coordination, rather than the four-fold coordination shown by other forms of SiO_2 and all of the common silicates. (The coordination of a cation is the number of oxygen atoms surrounding it. See *Earth Materials*, by W. G. Ernst, for further discussion of coordination and close-packed structures.) The structure and properties of stishovite are more closely analogous to the close-packed oxides than to the ordinary silicates.

The similarity between the extrapolated properties of the lower mantle and the properties of close-packed oxides was first pointed out by Birch. But his identification of oxide *structures* in the lower mantle does not necessarily imply that the *minerals* in this part of the Earth are simple oxides. Chemical compounds, including silicates, can exist, provided they have the required closeness of atomic packing. A further requirement demanded by the data seems to be that silicon must be in six-fold coordination.

The material of the lower mantle is undoubtedly very complex chemically, and since we have only two properties, φ and ρ, there is no hope of working out the composition in detail. On the basis of what is known of the abundances of elements in the solar system, we expect that the lower mantle is composed principally of the oxides of silicon and magnesium, with iron oxide making up about 10–20 per cent by weight. Lesser but significant amounts of CaO, Al_2O_3 and Na_2O are doubtlessly present, with other elements occurring in much smaller amounts. But our ignorance of the detailed chemical composition does not affect our conclusion about the occurrence of close-packed structures in this part of the Earth. Further information about the composition of the solar system will be found in *The Solar System*, by J. A. Wood.

Identification of the constitution of the lower mantle immediately reveals the nature of the transition zone. It is the region in which close atomic packing is achieved. The process takes place in more than one stage, as indicated by the steplike increases in velocity in the transition zone. Through a series of mineralogical transformations the close-packed state is attained.

Constitution of the Core

Years ago V. M. Goldschmidt suggested that the core is an iron-nickel alloy. His idea was based on the composition of the iron meteorites (see *The Solar System*, by J. A. Wood). We may test this hypothesis by the same extrapolations that we used to deduce the composition of the lower mantle. Values of ρ and φ for the core, extrapolated to zero pressure but corresponding to some unknown high temperature above the melting point of iron, are 6.2–6.8 gm/cm³ and 16–22 (km/sec)² respectively. The uncertainties are larger than in the case of the mantle because the range of pressure covered by the extrapolation is over three times as great. Pure liquid iron has a density of about 7 gm/cm³ and an elastic ratio of 19.4 (km/sec)². These figures suggest that ρ in the core is somewhat too small for liquid iron, and that φ is about the same. The density of an iron-nickel alloy is slightly higher than pure iron, and the data indicate that some lighter element is also present in the core in appreciable amounts. The possibility that silicon is an important constituent of the core has been argued by Ringwood.

The inner core is most plausibly interpreted as similar in composition to the outer core but solid in structure. Solidification could occur as indicated in Fig. 6–5. If the melting point increases with depth more rapidly than the actual temperature, the liquid outer core may exist outside of a solid inner core. The lower mantle is solid, on the other hand, because it has an intrinsically higher melting temperature than the core.

Alternatively, it has been suggested that the mantle and core are chemically identical and that further atomic rearrangements take place. Francis Birch has produced an apparently decisive argument against this possibility. Both φ and ρ can be measured in shock-wave experiments, and the seismic data yields the same quantities in the mantle and core. In Fig. 6–6 the "hydrodynamic velocity," which equals the square root of φ, is plotted against density for a number of elements and rocks, and for the mantle and core. Materials of different atomic number are well-separated on the plot. The mantle falls close to the curves for common rocks and for elements near aluminum in atomic number, whereas the core is close to iron. The figure

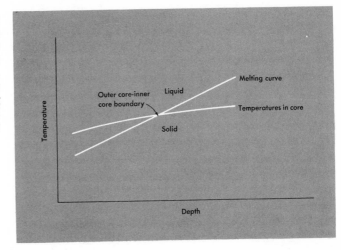

FIGURE 6-5 *If the melting temperature of the core material increases more rapidly with depth than the actual temperature, the actual temperature may intersect the melting curve, producing a solid inner core surrounded by a liquid outer core.*

The constitution of the Earth from seismic evidence

indicates that the mantle and core are chemically distinct, and the identification of the core as dominantly iron is confirmed.

This simple interpretation of the core will evidently require modification if, as seems likely, future seismic data force us to deal with a more complex structure. Further solid compounds, and possibly even immiscible liquids will have to be considered if the data demand. It seems most unlikely, however, that Birch's conclusion that the core is dominantly iron will be seriously affected by revisions of the seismic data.

FIGURE 6-6 $\varphi^{1/2}$ *versus density for elements, common rocks, and the mantle and core. Shaded area near the curves for the mantle represents the range of data found for shock compressions of common rocks. Data largely from F. Birch, 1961, Geophys. Jour., v. 4, p. 295.*

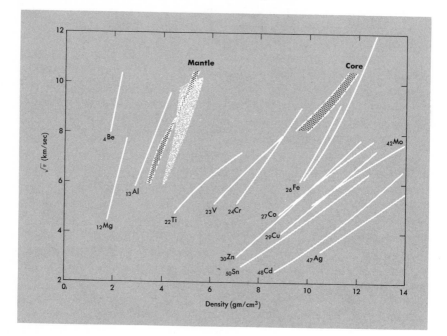

7

The Earth's gravity field

The original incentive for studying the Earth's gravity field arose from the fact that it provides a way of determining the shape of the Earth. It was found, however, that, in addition, measurements of gravity provide important information about the interior of the Earth and about mountains and ocean trenches. Previously, we have considered the Earth to be a perfect sphere, and this is very nearly the case. But we now must consider its small departures from sphericity. Because the departures are small, high accuracy is required in their measurement. As a consequence, *geodesy*, which is the science devoted to the determination of the shape of the Earth, has only recently come into its own. Since artificial satellites have been put in orbits close to the Earth, the first geodetic observations of really high accuracy can be made. The improvement in our geodetic knowledge has been one of the major fruits of space research.

It is important at the outset to define precisely what geodesists mean by the shape of the Earth. They do not refer to the actual surface of the solid Earth with its complex irregularities, but instead to a fictitious surface, called the *geoid*, which is an *equipotential*, or *level*, surface. The normal to the geoid points in the direction of the vertical everywhere, that is, in the direction of the local force of gravity. The orientation of the level surface can be measured with a plumb bob, which defines the direction of the normal, or by a spirit level, which gives the orientation of lines lying in the horizontal surface. Since gravity is normal to the level surfaces, there is no component of force along them. Therefore, the surface of a liquid at rest will be level. It is important to note that the *force* of gravity is not necessarily constant over a level surface. Instead, the surface is defined in terms of the *direction* of gravity.

The Earth is surrounded by an infinite number of level surfaces at various distances from its center. The geoid is defined as that particular level surface that coincides with mean sea level at sea. The geoid can be projected beneath the land with the help of gravitational theory, but some assumption about the density of the rock between the geoid and the surface must be made. Different assumptions lead to slightly different positions of the geoid.

Sir Isaac Newton reasoned that the centrifugal force arising from the Earth's rotation would cause the equator to bulge outward and the polar regions to be flattened. A. O. Clairaut showed that the Earth's shape should be nearly that of an oblate spheroid, provided it had negligible strengths in its interior. Much of the effort of geodesists has gone towards determining the dimensions of the spheroid that most nearly coincides with the more complexly shaped geoid. This geometrical figure, known simply as *the spheroid*, is the most important standard of reference in geodesy. It approximates the geoid much more closely than could a sphere because it takes account of the equatorial bulge, which is the main departure from sphericity exhibited by the Earth.

Gravity Anomalies

We recall that a geophysical anomaly, as defined in Chapter 3, is found by subtracting the value of some geophysical quantity, as given by a reference standard, from the value actually measured. The principal source of variation of gravity over the Earth's surface arises from the equatorial bulge and the centrifugal force due to rotation. They combine to produce a change in gravity from equator to pole that is about 10 times as large as any other variation. Subtracting this major change emphasizes smaller features that we seek to study.

Gravity is defined as the force of attraction felt by a body attached to the Earth and rotating with it, divided by the mass of the body. Hence it is measured in units of acceleration, that is, cm/sec^2. The unit is the gal, named for Galileo, and equal to 1 cm/sec^2. The Earth's gravitational field is about 980 gals, and an accuracy of measurement of better than 1 milligal (one-thousandth of a gal and approximately one-millionth of the total field) is commonly achieved at land stations. Gravity can also be measured on ships at sea, but the results are affected by accelerations produced by waves, and the accuracy is reduced to 5 or 10 milligals.

The standard of reference used in gravity is the gravitational field of the spheroid. In this case, gravity depends only on latitude and is given by the International Gravity Formula,

$$g_t = 978.049 \ (1 + 0.0052884 \sin^2 \theta - 0.0000059 \sin^2 2\theta),$$

where θ is latitude and g_t is given in gals. The constants in this formula depend on old measurements. The equation does not represent the gravity field of the spheroid that most closely fits the geoid according to the most recent data, but it is an amply accurate standard nevertheless.

The International Gravity Formula gives the standard value of gravity at the level of the spheroid. A precise definition of a gravity anomaly would be the difference between the standard value on the *geoid* and the value of gravity actually measured at the same point. In practice, the difference in level between spheroid and geoid is usually neglected, and values at the differents levels are compared directly. On land, however, gravity is measured at the surface, which is commonly far above the geoid, and we must make corrections to the measured value in order to obtain the value that would be observed at the level of the geoid. The corrections are made in three stages. First we add to the measured value the increase in gravity due to the fact that the geoid is closer to the center of mass of the Earth than is the point of observation. This correction is analogous to the difference in gravity that we would observe if we made one measurement high in the air (assuming that this could be done) and another on the ground directly beneath. It is called the *free-air correction,* and if we were to subtract the standard value of gravity at this stage in the calculation, we would be left with a *free-air anomaly.* In fact, there is rock rather than air between the point of observation and the geoid, and we must subtract its gravitational attraction. As a first approximation, we treat the rock as a layer having uniform thickness equal to the distance between the measuring point and the geoid. This leads to the *simple Bouguer correction* and to the corresponding *simple Bouguer anomaly* upon subtraction of standard gravity. But, in general, there are hills projecting above the elevation of the observation and valleys extending below. Their gravitational effects add to each other, and they represent the final corrections that must be added to measured gravity to give *complete Bouguer corrections* and *complete Bouguer anomalies.* The simple Bouguer corrections and anomalies differ little from their complete Bouguer analogues except in regions of rough topography, and differences exceeding a few tens of milligals are rare.

Somewhat analogous corrections must be made to gravity observations at sea. Early measurements were made in submarines and required both free-air and Bouguer corrections. Modern measurements are made in surface ships and require no free-air correction since they are made at sea level, which is the level of the geoid. The Bouguer correction is added to the observations to account for the change in gravity that would result if the sea were filled with rock instead of water.

The various corrections and corresponding anomalies for a gravity station at a fairly high elevation is given in Fig. 7–1. The difference between simple

Blind Bull, Wyoming
Elevation: 2492.8 m.
Gravity values in milligals

Theoretical at sea level	980,443.9	Observed	979,722
Free air correction	− 769.2		− 979,675
	979,674.7	Free air anomaly	+ 47
Simple Bouguer correction	+ 278.9		
	979,953.6		979,722
Terrain correction	− 2.0		− 979,954
	979,951.6	Simple Bouguer anomaly	− 232
			979,722
			− 979,952
		Complete Bouguer anomaly	− 230

FIGURE 7-1   *The various corrections to gravity data and the resultant anomalies for a station in Wyoming.*

and complete Bouguer corrections is small, and the Bouguer anomalies are much larger in magnitude than the free-air anomaly and are negative. In high places Bouguer anomalies prove to be characteristically negative and typically have magnitudes of a few hundred milligals. Free-air anomalies, on the other hand, are generally much smaller in magnitude, can have either sign, and the average value over a mountain chain can be near zero. In the oceans Bouguer anomalies are positive with magnitudes of 200–300 milligals. Free-air anomalies are again smaller and may average nearly zero over large regions. The interpretation of these results is the subject of the next section.

The Theory of Isostasy

The observational evidence is that free-air anomalies are relatively small and in a given region tend to be negative about as often as they are positive. Bouguer anomalies, on the other hand, are almost always negative in mountains and high plateaus and positive over the deep sea. Furthermore, the magnitude of the Bouguer anomaly is related to the elevation of the land or to the depth of the water. These observations suggest that the Bouguer corrections are too large. The excess material between the Earth's surface and the geoid on land, and the deficient material represented by water rather than rock at sea, do not have as large a gravitational effect as is predicted from their masses and dimensions.

The first observation that mountains have unexpectedly small gravitational attractions was made by Pierre Bouguer in about 1750. Bouguer's result was confirmed by Sir George Everest's survey of India about 100 years later. The discrepancy between the expected attraction and the attraction actually observed was noted by Archdeacon Pratt in England, and he offered the following explanation. Mountains are not simply masses lying on the Earth's surface. Instead, they are produced by abnormally high temperatures in the Earth's interior, which cause the material to expand and therefore to decrease in density. The outward bulge of the mountains is a consequence of the expansion, and no extra mass is involved. The mass of material per unit area between the surface and some level, called the *depth of compensation*, is the same everywhere.

Pratt's theory was criticized by Sir G. B. Airy, who offered an alternative explanation. In his view, the Earth's crust floats on a weak but not necessarily liquid substratum of higher density. In mountains, where the surface is high, the light crust is abnormally thick and projects deeply into the substratum in the same manner that an iceberg projects into the sea. Airy's hypothesis does not lead to a single depth of compensation in the way that Pratt's explanation does, but the mass per unit area between the surface and the level of the deepest projection of the crust is everywhere the same. Since the crust is assumed to thicken under mountains, Airy's scheme of compensation is sometimes called the *roots of mountains* hypothesis. The notion that mountains are not excess loads placed on the surface, but rather that their visible mass is counterbalanced by a deficiency of mass at depth, is termed the *theory of isostasy*, and the compensation of excess mass at the surface by a reduction in mass at depth is termed *isostatic compensation*. No strength is required of the material below the depth of compensation. This material makes up the asthenosphere (see Chapter 4), and the concept originated in the theory of isostasy many years before seafloor spreading was proposed.

Isostatic compensation can be achieved in an infinite number of ways, and the hypotheses of Pratt and Airy represent two of the possibilities. They are illustrated in Fig. 7–2, in which, for simplicity, the crust is treated as a number of disconnected prisms floating in a denser fluid substratum. Pratt's hypothesis was used extensively in the United States by J. F. Hayford and has become known as the Pratt-Hayford scheme of compensation. Similarly,

Airy's hypothesis was adopted in Europe by W. A. Heiskanen and is known as the Airy-Heiskanen scheme.

The quantitative calculation of isostatic compensation is easily made from the condition of constant mass per unit area above some specified level. Calculation for the Pratt-Hayford scheme is shown in Fig. 7–3 and for the Airy-Heiskanen scheme in Fig. 7–4. Note that in the latter case the thickness of the sea-level crust does not enter the equations expressing the conditions for compensation. Each scheme provides a way of calculating the distribution of mass in the Earth's interior from surface elevations. Gravitational theory then enables us to calculate the attractions of the compensating masses and to make a further isostatic correction to observed gravity, leading to an isostatic *anomaly*.

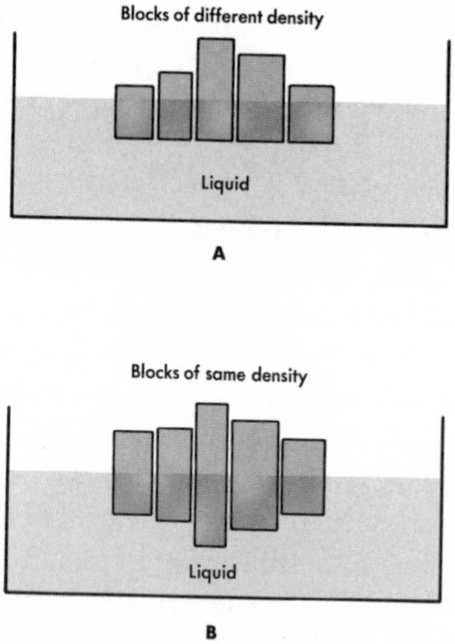

FIGURE 7-2 *Schematic illustration of (A) the Pratt-Hayford and (B) the Airy-Heiskanen schemes of isostatic compensation.*

Both schemes of isostatic compensation contain a parameter that can be adjusted to minimize the anomalies. In the Pratt-Hayford scheme, it is the depth of compensation; in the Airy-Heiskanen, it is the thickness of the crust at sea level. (Although the latter quantity does not appear in the expression for the thickness of the mountain root, the gravitational attraction of the compensating masses does depend on it.) These parameters are adjusted until the resulting isostatic anomalies are minimized over an area of continental dimensions, such as Europe or the United States. Because they provide an adjustable parameter, both the Pratt-Hayford and Airy-Heiskanen anomalies are usually somewhat lower than the free-air anomalies. But gravity data are of little help in choosing between the two schemes of compensation because they are about equally successful in reducing anomalies.

Alternate lines of geophysical evidence do not help to choose between the Pratt-Hayford and Airy-Heiskanen schemes of compensation either, but they indicate why it is difficult to reach a clear-cut decision. The thickness of the crust at sea level that leads to the smallest anomalies on the Airy-Heiskanen scheme is about 50 km. This is fairly close to the depth to the Moho in continental regions. The Pratt-Hayford scheme yields the smallest anomalies if the depth of compensation is somewhat greater than 100 km, which places it near the top of the low-velocity zone in continental regions.

The Earth's gravity field

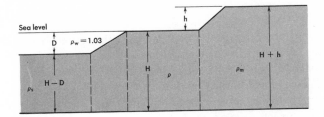

For isostatic balance,

$$H\rho = (H - D)\rho_s + D\rho_w = (H + h)\rho_m.$$

Whence,

$$\rho_s = \frac{H\rho - D \times 1.03}{H - D}$$

$$\rho_m = \frac{H\rho}{H + h}$$

FIGURE 7-3 *The calculation of isostatic balance on the Pratt-Hayford scheme.*

A major change in density takes place at the Moho, and isostatic compensation of the Airy-Heiskanen type will result from changes in the thickness of the seismically defined crust. This factor dominates the difference in

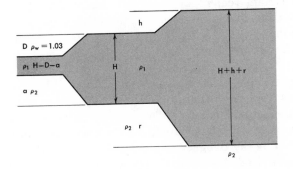

For isostatic balance,

$$\begin{aligned} H\rho_1 + r\rho_2 &= (H - D - a)\rho_1 + D\rho_w + (a + r)\rho_2 \\ &= (H + h + r)\rho_1. \end{aligned}$$

Whence,

$$a = \frac{D(\rho_1 - 1.03)}{\rho_2 - \rho_1}$$

$$r = \frac{h\rho_1}{\rho_2 - \rho_1}$$

FIGURE 7-4 *The calculation of isostatic balance on the Airy-Heiskanen scheme.*

The Earth's gravity field

structure beneath continents and oceans. On the other hand, the Pratt-Hayford depth of compensation corresponds more closely to the depth to the asthenosphere if the latter coincides with the low-velocity zone. A combination of Pratt-Hayford and Airy-Heiskanen compensation may be closer to the truth than either scheme by itself. Flow in the asthenosphere may be required to correct imbalances that are not removed by changes in crustal thickness. The process of compensation is doubtlessly complicated, and there is no reason to suppose that it is accomplished in the same manner everywhere.

The theory of isostasy offers a ready explanation of the deep erosion that has affected ancient mountain systems. Erosion removes mass from above the geoid; consequently, the mountains become overcompensated. The compensating mass is lighter than its surroundings and therefore is buoyant, so it tends to rise. The total vertical movement may be many times the height of the mountains. The common exposure of the deeply eroded cores of ancient mountains is a direct consequence of this process of isostatic uplift.

Departures from Isostatic Compensation

There are certain regions of the Earth where isostatic anomalies are comparatively large. Stable continental regions that have not been subjected to orogeny in the last several hundred million years commonly have gravity anomalies that vary by about 100 milligals yet show no particular relation to topography. A striking example of such a feature is provided by the mid-continent gravity high in the central United States (Fig. 7–5). This sharp feature extends from the western tip of Lake Superior to central Oklahoma. This part of the country is almost flat. As a consequence, isostatic corrections would be nearly constant throughout the area shown in the figure, and the range shown by the Bouguer anomalies—from −70 to +60 milligals approximately—would not be changed appreciably by allowance for isostasy. The gravity high is believed to be caused by a thick accumulation of rock that is denser than the surrounding crustal material. The feature may be as much as one billion years old, and it must be several hundred million years old. There is no evidence that the excess mass is in any way compensated, and the pattern of anomalies indicates that any compensation that might exist must be located at considerable depth. The excess mass causing the anomaly must be supported by the strength of the lithosphere, and the support has apparently persisted for a long time, long even from the geologic viewpoint. There is independent evidence that the rocks may be cooler than normal beneath this part of the world (see Chapter 8), which is consistent with unusually great strength.

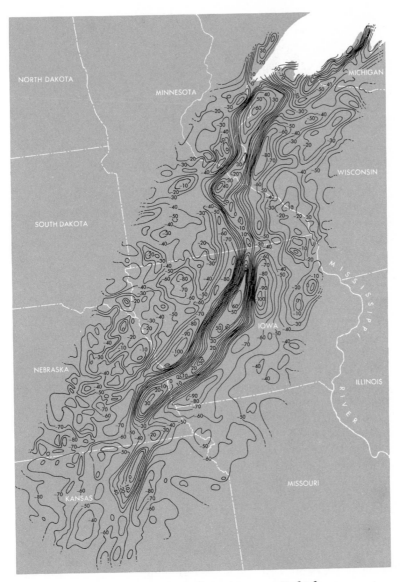

FIGURE 7-5 *The mid-continent gravity high.*

Other regions where lack of isostatic compensation is even more promi-
nently displayed are the deep ocean trenches. The depth of water can exceed 10 km, and the isostatic anomalies, which are always negative, can
exceed 200 milligals. The strongly negative anomalies, which occur in elon-

The Earth's gravity field

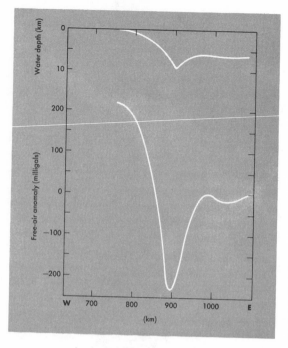

FIGURE 7-6 *Free - air anomalies across the Tonga trench south of Fiji. Data from Talwani, Worzel, and Ewing, 1961,* Jour. Geophys. Res., v. 66, p. 1265.

gate belts shaped rather like the trenches themselves, were discovered by F. A. Vening Meinesz (see Fig. 7–6). The explanation of these isostatic anomalies is entirely different from the one advanced to account for the anomalies in stable continental regions. Strength plays a comparatively minor role; the major one is played by the tongue of material descending along the Benioff zone. It moves rapidly enough to drag the surrounding mantle with it, producing the topographic depression of the trench and the isostatic anomalies. Isostatic compensation is not achieved fast enough to prevent the anomalies, which thus owe their origin to the dynamic process of sea-floor spreading.

The Shape of the Earth

For many years the best way of determining the shape of the Earth, by which we mean the shape of the geoid, was to calculate it from observations of gravity.But gravity is affected by local anomalies, for which precise allowance cannot be made because isostatic compensation does not occur to a perfect, predictable extent. A further difficulty arises because there remain large areas, particularly at high latitudes, in which gravity has never been measured. These difficulties do not affect the tracking of artificial satellites, and the size and shape of the Earth are best determined toray from satellite orbits.

Satellites move in the Earth's gravity field, and their orbits, when accurately determined, give important information about that field. Observations of the satellites enable us to determine the shape of a level surface at an altitude of a few hundred kilometers. By gravitational theory we can project this surface downwards without ambiguity until it encounters matter. Some assumption about the density must be made if downward projection is to be continued, and the simplest is to neglect the mass of material above the geoid. This is completely analogous to the calculation of a free-air anomaly, and the geoid obtained by projecting the satellite data to sea level in this way may be termed the free-air geoid. This choice of treatment of the mass above the geoid is strongly dictated by convenience, but it is defensible in view of the observation that free-air anomalies are not much larger than those derived from isostatic hypotheses.

It is easiest to depict the geoid with reference to the spheroid that matches it most closely. The size and shape of a spheroid can be specified by giving its equatorial radius, a, and polar radius, c, but another convention is adopted in geodesy. In addition to the equatorial radius, the *coefficient of flattening*, f, defined as $(a - c)/a$, is used. In terms of f the polar radius is then $c = (1 - f)a$. The value of f determined from satellite data is about $\frac{1}{298}$. This is slightly larger than would be found if the Earth were completely devoid of strength. In that case, the "equilibrium" value of f is estimated to be close to $\frac{1}{300}$.

Satellite geodesy has revealed other departures of the geoid from the spheroid that are comparable in size to the departure of the coefficient of flattening from its equilibrium value. They too are evidence of strength in the Earth. The Earth is "pear-shaped," in that at a given latitude the mean radius in the southern hemisphere is slightly larger than in the northern hemisphere. The difference in radii amounts to about 15 m in middle latitudes. Contours showing the difference in height between the geoid and spheroid are shown in Fig. 7–7. In the Indian Ocean a pronounced depression of the geoid relative to the spheroid is prominent, as is a ridge running from Europe through eastern Africa. Both humps and hollows occur in the Pacific. The undulations of the geoid show little or no correlation with continents and oceans.

The locations of the anomalous masses giving rise to the undulations of the geoid cannot be specified. They may be the result of lateral variations in temperature in the upper mantle, unrelated to continentality. Alternatively, they may reflect lateral inhomogeneity supported by strength in the lower mantle. We cannot be sure. The difference between the observed flattening of the spheroid and the equilibrium value is similarly of uncertain origin. The excess bulge could be left over from a time in the past when the Earth was rotating more rapidly, but since this departure is no larger than the other undulations of the geoid, there is no compelling reason to give a special explanation for it.

The Earth's gravity field

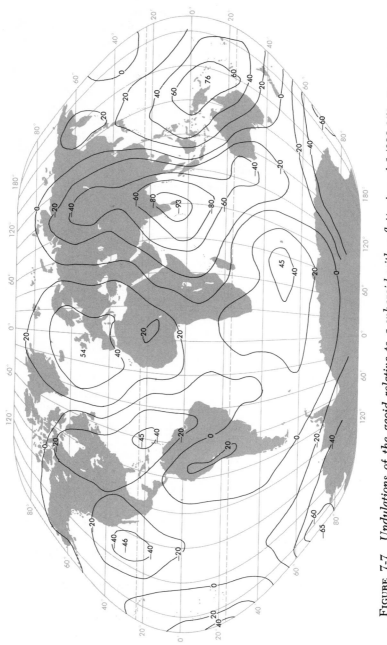

FIGURE 7-7 *Undulations of the geoid relative to a spheroid with a flattening of 1/298.252. Data from E. M. Gaposchkin, 1967, Space Res., v. 7, p. 685.*

Once the coefficient of flattening is known, it can be combined with astronimical measurements of the precession of the equinoxes to give the Earth's moment of inertia. Precession of the equinoxes is analogous to the motion of a top that is set spinning at an angle to the vertical, as shown in Fig. 7–8. Gravity produces a torque on the top tending to upset it, but this tendency is opposed by the top's rotation. It remains upright, and its axis describes a cone about the vertical, as indicated in the figure. This motion is called *precession*. A similar torque is experienced by the Earth, as is also indicated in Fig. 7–8. The equator is inclined to the planes of the Earth's orbit about the Sun and of the Moon's orbit about the Earth. The gravitational attraction of the Sun and Moon tends to twist the equatorial bulge into the planes of their orbits, but since the Earth is rotating, the effect of this torque is to cause precession of the axis of rotation, as in the case of the top. The expression relating the rate of precession to the geometrical shape of the Earth contains the Earth's moment of inertia as well. Once the coefficient of flattening and the precessional constant are known, the equation can be used to calculate the moment of inertia. The mass of the Earth is another quantity found from geodetic observations. It is calculated from observed gravity at the surface and the known radius of the Earth.

FIGURE   7-8 *Torques causing precession of a top and of the Earth.*

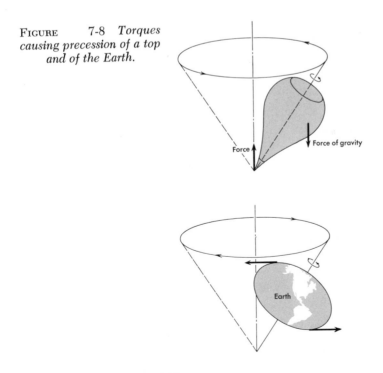

The Earth's gravity field

117

8

Heat flow and the temperatures

in the Earth

We saw in Chapter 6 that the pressure distribution in the Earth is reasonably well-determined. There is no relatively simple, direct way of estimating temperatures at great depths, analogous to the way we estimated the pressures previously. Direct observations of temperature are confined to the depths reached by drilling, and these depths are less than 10 km. Extrapolation of temperature downwards in the Earth requires knowledge of quantities that we cannot estimate precisely; consequently, the temperatures are only roughly known at great depths.

Despite the difficulty in determining them accurately, it is important to know the temperatures in the Earth. The properties of the interior that we can determine, such as the seismic velocities and density, depend on temperature as well as on pressure and composition. In the deep interior, temperature affects these properties less than pressure does, but it is still necessary to have good estimates of temperature if we are to infer the chemical composition and mineralogical constitution with high accuracy from seismic data. In addition, the temperatures are, in part, the products of events that took place in the past, unlike the gravity field, say, which depends only on the present distribution of mass. It is even conceivable that thermal events accompanying the formation of the Earth have left a recognizable imprint on the internal temperatures today. If we can determine the temperatures, it may help us deduce the way in which the Earth originated and will certainly improve our knowledge of the composition and constitution of its interior.

Terrestrial Heat Flow

Two thermal quantities that we can measure at the surface are the mean annual temperature and its rate of increase with depth. The latter quantity is called the *geothermal gradient*. Thermal energy flows down a thermal gradient just as electric current flows in response to a voltage drop, and the amount of energy flowing through unit area in unit time is called the *heat flow*. We shall measure heat flow in heat-flow units (abbreviated HFU); 1 HFU equals 10^{-6} cal/cm^2 sec. Heat flow is proportional to the thermal gradient, and the constant of proportionality is called the *thermal conductivity*. Thermal conductivity times thermal gradient equals heat flow. The thermal conductivity of rocks can readily be measured, and observations of geothermal gradients can easily be converted to measurements of terrestrial heat flow once the thermal conductivity is known. It is preferable to work with heat flow rather than thermal gradient because the former is a flow of energy. Energy is released in the Earth by heat sources, which manifest themselves directly as heat flow and only secondarily as a thermal gradient.

The worldwide average heat flow is 1.5 HFU, and this heat is radiated into space. The heat flow is small, being sufficient to melt a sheet of ice only one-half centimeter in thickness in a year. It represents about one-ten-thousandth of the heat received from the Sun during the day and reradiated during the night, so its effect on the temperature at the surface is utterly negligible. But compared to other manifestations of the Earth's internal heat, the total amount of heat flowing through the surface is a relatively large number. It is estimated to be ten to one hundred times as great as the heat lost through volcanic activity, for example.

The measurement of terrestrial heat flow is simple in principle, but it requires access to undisturbed temperatures beneath the surface. Oil wells, mines, tunnels, and holes drilled for mineral exploration have all afforded sites at which the geothermal gradient has been determined. The requirement that the temperatures be undisturbed sets limits to the minimum depth from which useful measurements can be obtained. On land, disturbances are caused by annual and daily temperature fluctuations, circulating ground water, climatic variations, and uplift and erosion. Some of them can be avoided by making measurements at fairly great depths. The effects of climatic variations and of uplift and erosion cannot be escaped in this way, and the measurements must be corrected when these sources of disturbance occur. In practice, it is found that useful data can be obtained if access to depths greater than about 200 m is available.

On the sea floor near-surface disturbances are much less, and the gradients are commonly measured within a few meters of the surface by using temperature sensors mounted on coring tubes (see *Oceans*, by K. K. Turekian, for a discussion of coring at sea). Reliable results could not possibly be obtained in this way on land, and they are possible at sea only if the temperature of the bottom water has remained constant for the past few hundred years. This is apparently the case in deep water, but geothermal gradients at sea are sometimes disturbed by rapid sedimentation and submarine slumping.

Geothermal gradients are multiplied by the thermal conductivity of the rocks or sediments in which they are observed to give the heat flow. The thermal conductivity of common types of rock at room temperature lies in the range 0.003 to 0.010 cal/cm sec°C, and ocean sediments range from 0.002 to 0.003 cal/cm sec°C. The geothermal gradients required to give a normal heat flow of 1.5 HFU with this range of conductivities vary from 15 to 75°C/km.

Observations of terrestrial heat flow are shown in Fig. 8–1, where the number of measurements falling in a given interval is plotted against the heat flow in that interval. The great majority of measurements falls in the range of 1–2 HFU; the value most frequently found is 1.3 HFU. Most values

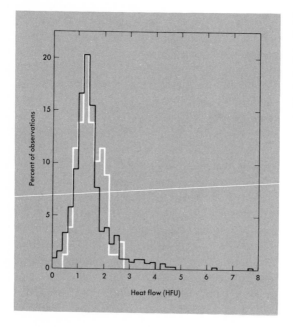

FIGURE 8-1 *Distribution of heat flow observations. Black curve represents oceanic determinations, and white curve represents continental data. Data from von Herzen and Lee, 1969, Am. Geophys. Un., Monograph 13, p. 88.*

below 0.6 HFU were measured in the oceans and probably are affected by some disturbance such as slumping. On the high end of the range, the values greater than 2.0–2.5 HFU represent thermal areas where hot magma or hot water has recently risen towards the surface. In many of these places there are hot springs or recent lavas that give cororborative evidence of recent thermal activity, but in some land areas and at most undersea stations where high heat flow is found, the heat flow itself is the principal evidence of recent thermal activity.

Although the heat flow is still unknown over about half of the Earth's surface, we have enough data to draw some general conclusions about its behavior. Low and rather uniform values of flux are found in the deep ocean basins, the trenches, and stable continental regions that have been inactive geologically for several hundred million years. The heat flow is very variable but high on the ocean ridges, as would be predicted from the hypothesis of sea-floor spreading. The heat flow can be very high in seas that are thought to represent recent rifting and fragmentation of continents, such as the Red Sea–Gulf of Aden depression or the Gulf of California. On land, heat-flow provinces can be recognized. In North America there is an eastern province with heat flow averaging slightly more than 1 HFU and ranging from 0.7–2.3 HFU and a Cordilleran province with higher (averaging about 2 HFU) and more variable heat flow. There may be a third province of low to normal heat flow along the Pacific Coast, but present data are sufficient to define it only in the Sierra Nevada and in Southern California. Within each province the principal variation in heat flow is attributable to change in radioactive heat production in near-surface rocks. The different provinces apparently are the result of variations in heat flow from the mantle. Heat-flow provinces are discussed in the next section.

The heat-flow pattern in Japan is an interesting illustration of its variation near a trench and its associated volcanic island arc. A large variation in heat flow takes place across the islands (Fig. 8–2), with values in excess of 2 HFU occurring from the line of active volcanoes westward into the Sea of Japan and southward along the Izu-Bonin arc. Along the eastern margin of the islands the heat flow drops to about 1 HFU, and even lower values are found in the trench to the East. The low heat-flow area is due to the tongue of cool oceanic lithosphere descending along the Benioff zone, while the high heat-flow area is supported by thermal activity, of which the line of volcanoes down the center of the Japanese islands is the most dramatic manifestation.

Heat Generation by Radioactivity

The amount of heat produced by the different radioactive elements is controlled by their abundances, by their rates of decay, and by the amount

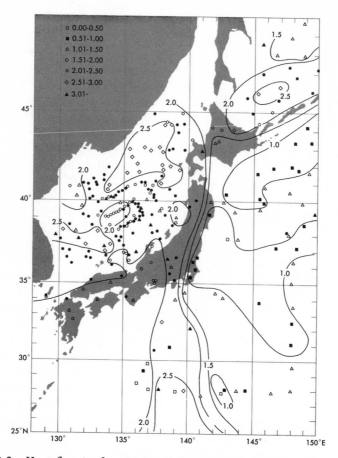

FIGURE 8-2 *Heat flow in the vicinity of the Japanese Islands. Data from Horai and Uyeda, 1969*, Am. Geophys. Un., Monograph 13, p. 95.

of heat released in each disintegration. The rate of decay is inversely proportional to the half-life of an isotope, which is defined as the time required for half of the atoms originally present to disintegrate. Only isotopes with half-lives of about 10^9–10^{10} years are presently important to temperatures in the Earth. Those with much shorter half-lives have decayed to the point where their abundances are very small, and those with much longer half-lives disintegrate too slowly to release significant amounts of heat. Only the two isotopes of uranium, U^{235} and U^{238}, the isotope of thorium, Th^{232}, and a rare isotope of potassium, K^{40}, are of geothermal importance.

Determination of the concentrations of the important radioactive elements in rocks has been a difficult problem in analytical chemistry ever since its significance was first appreciated in the early part of this century. Uranium and thorium are trace elements commonly present in amounts less than one part per million, and great care must be taken to avoid contamination of the sample during analysis. The chance of sizable error increases as the concentrations of these elements decrease. Potassium is normally regarded as a major element, but in many important types of rock it is found in amounts that do not exceed a few tenths of one per cent. Standard analytical methods are often inadequate at these low concentrations, and the error is usually to overstate the potassium content. The uranium and thorium contents are also overstated when precautions against contamination are insufficient. Analytical techniques have been greatly improved in recent years, and it is believed that reliable data are now being obtained.

Radioactive heat production in rocks shows a definite association with rock type. Rocks that occur commonly at the Earth's surface characteristically have 10 to 1000 times greater contents of radioactive elements than rocks with seismic velocities appropriate to the mantle. Additional information about the concentration of heat production towards the surface comes from further study of the heat-flow provinces that we mentioned in the last section. Two examples of the relation between heat flow and heat production in the surrounding rocks are shown in Fig. 8–3, one for the northeastern United States and the other for the Sierra Nevada. In both cases the heat

FIGURE 8-3 *Heat flow versus local heat generation in the Sierra Nevada and the northeastern United States. Data from Birch, Roy, and Decker, 1968,* Studies of Appalachian Geology, *ed. Zen, White, Hadley, and Thompson, Interscience, p. 437, and from Lachenbruch, 1968,* Jour. Geophys. Res., *v. 73, p. 6977.*

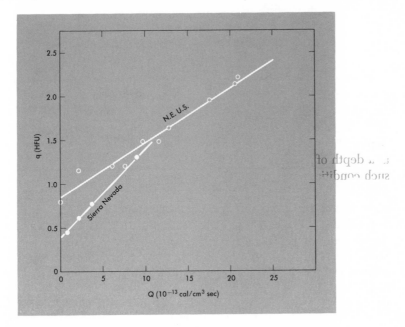

Heat flow and the temperatures in the Earth

flow is a linear function of local heat production; that is, the heat flow, q, is related to the heat production, Q, by the equation $q = q_0 + Qd$, where q_0 and d are constants. This relation can be interpreted simply by assuming that Q does not change with depth in a near-surface layer of constant thickness. The constant d then represents the thickness of the layer, and q_0 is the amount of heat flowing through it from below. We can calculate d from the slopes of the lines in Fig. 8–3, and we can obtain q_0 from the intercepts at $Q = 0$. The results are $d = 6$ km and $q_0 = 0.9$ HFU in the northeastern states, and $d = 10$ km and $q_0 = 0.4$ HFU in the Sierra Nevada. We need not take Q to be constant in the radioactive layer, but other simple assumptions about its variation with depth lead to the same conclusion. Material having the radioactivity observed at the surface is no more than about 10 km thick; the remainder of the crust and the underlying mantle are relatively low in radioactivity.

Heat-flow provinces are defined by straight lines on a plot of q versus Q. Measurements falling on the same line are in the same province, and those falling on different lines are in different provinces. As yet, we hardly have enough information to deduce the nature of the transition zone between one heat-flow province and another, but the data we have suggest that the boundaries of the provinces are sharp. The different provinces seem to represent segments of the crust and upper mantle that evolved separately. Separation in the time of evolution is common, and different parts of what is now one continent may have developed in separated regions, to be united by the process of plate tectonics.

Temperatures Deep in the Earth

Studies of heat flow provide important information about near-surface temperatures, but the extrapolation of the results to great depths is far from straightforward. The geothermal gradient is typically found to be about 20°C/km near the surface, but it must decrease markedly with increasing depth. Otherwise, the temperature would be in the neighborhood of 2000°C at a depth of 100 km, which is only about $\frac{1}{60}$ the way to the center. Under such conditions, any plausible mantle rock would be wholly molten and unable to transmit S-waves. However, seismic data show that this is not the case; hence, the geothermal gradient must decline with depth.

Prediction of the temperature as a function of depth from heat-flow theory requires knowledge that we do not have. We would need both detailed information about the distribution with depth of radioactive heat sources and details of the mechanisms of heat transfer. Ordinary conduction is not only the means by which heat is transferred at the high temperatures inside the Earth. Heat radiation plays an important role, and the motions of material implied by sea-floor spreading may dominate the thermal regime at depths greater than a few tens of kilometers. Uncertainties introduced by our lack of accurate information about these processes prevent the calculation of temperatures from heat-flow data.

There are two other ways of estimating temperatures in the Earth that are independent of heat-flow observations and are capable of yielding useful results. One method involves the seismic velocities, which increase with depth because of the increasing pressure. But the rate of increase in velocity depends in part on the geothermal gradient, decreasing as the rise in temperature becomes larger. Birch has estimated the geothermal gradient that is compatible with the observed velocity gradients in the lower mantle, under the assumption that the material there is homogeneous. He could only conclude that a temperature of 5000°C in the deep interior was compatible with the velocity gradients, whereas a temperature of 10,000°C seemed to be ruled out. However, as our knowledge of the behavior of matter at the high pressures and temperatures deep in the Earth improves along with our knowledge of the velocity and density distributions, this approach to the temperature problem is likely to yield much more accurate results.

A second way of determining the temperatures at certain depths in the Earth is afforded by the mineralogical transformations that were discussed in Chapter 6. The first problem is to identify the transformations responsible for the observed seismic discontinuities. The temperatures of the transformations must then be determined at the pressures corresponding to the depths at which the discontinuities are observed. This is done either by direct experimentation in the laboratory or by theoretical calculation. For example, Ringwood has identified the cause of the jump in velocities at a depth of about 400 km as an atomic rearrangement through which the mineral olivine acquires the atomic structure of spinel, $MgAl_2O_4$. Furthermore, he and his collaborators have observed the transformation in the laboratory. If their interpretation of the seismic data is correct, the laboratory data indicate a temperature of 1500°C–1600°C at a depth of about 400 km.

In Chapter 6 we discussed the possibility that partial melting is responsible for the low-velocity zone. If we assume this to be the case, and assuming further that peridotite is melting in the presence of water, then available laboratory data indicate that the temperature is perhaps 900°C–1000°C at the top of the low-velocity zone. This estimate, of course, is no better than the assumptions and data on which it is based.

Another speculative estimate of temperature, this time at much greater depths, is provided by the boundary between the outer and inner cores. If this seismic break is due to melting, and if the core is dominantly iron, then the temperature at this depth is estimated to be between 3000°C and 4000°C. The adoption of a more complex structure for the core, which we suspect will be required, will necessitate modification of these figures. The size of the required revision is not clear at present, but the temperatures as they stand are at least in agreement with Birch's conclusion, derived from studies of the seismic velocities, that the maximum temperature in the Earth is much closer to 5000°C than 10,000°C. It is unfortunate that we must conclude on such an uncertain note, but much remains to be done before temperatures in the Earth can be accurately known.

Suggestions for further reading

General

Tucker, Cook, Iyer, and Stacey, *Global geophysics.* New York: American Elsevier, 1970.

Stacey, F. D., *Physics of the Earth.* New York: Wiley, 1969.

Hart, P. J., ed., *The Earth's crust and upper mantle.* Geophysical Monograph 13. Washington: American Geophysical Union, 1969.

Chapter 2
Geologic structures

Shelton, J. S., *Geology illustrated.* San Francisco: Freeman, 1966.

Chapter 3
The Earth's magnetic field

Irving, E., *Paleomagnetism and its application to geological and geophysical problems.* New York: Wiley, 1964.

Chapter 4
Plate tectonics

Takeuchi, Uyeda, and Kanamori, *Debate about the Earth.* San Francisco: Freeman, Cooper, 1967.

Isacks, Oliver, and Sykes, Seismology and the new global tectonics. *Jour. Geophys. Res.,* v. 73, pp. 5855–5899, 1968.

Morgan, W. J., Rises, trenches, great faults, and crustal blocks. *Jour. Geophys. Res.,* v. 73, pp. 1959–1982, 1969.

Chapter 5
Seismology

Bullen, K. E., *An introduction to the theory of seismology.* Cambridge University Press, 1963.

Hodgson, J. H., *Earthquakes and Earth structure.* Englewood Cliffs, N.J.: Prentice-Hall, 1964.

Chapter 6
The constitution of the Earth from seismic evidence

Anderson, Sammis, and Jordan, Composition and evolution of the mantle and core. *Science,* v. 171, pp. 1103–1112, 1971.

Ringwood, A. E., Phase transformations and the constitution of the mantle. *Phys. Earth and Plan. Interiors,* v. 3, pp. 109–155, 1970.

Chapter 7
The Earth's gravity field

Heiskanen, W. A. and Vening, Meinesz, F. A., *The Earth and its gravity field.* New York: McGraw-Hill, 1958.

Chapter 8
Heat flow and the temperatures in the Earth

Lee, W. H. K. ed., *Terrestrial heat flow.* Geophysical Monograph 8. Washington: American Geophysical Union, 1965.

Index

Airy, Sir G. B., 109, 110
Airy-Heiskanen scheme,
    110–112
Anderson, D. L., 91
Anelasticity, 90
Anomalies
    Bouguer
        complete, 107
        simple, 107
    defined, 39–40
    free-air, 107
    gravity, 106–108
Asthenosphere, 65

Basalts, 65–66
Bedding, 10–11
Benioff, Hugo, 56
Benioff zones, 56–57
Birch, Francis, 97–98
Bloom, A. L., 1
Body waves, 69
    behavior at an interface,
        77–83
    travel times of, 73–77
Bouguer, Pierre, 109
Bouguer anomaly
    complete, 107
    simple, 107
Bouguer correction
    complete, 107
    simple, 107
Bullard, E. C., 28–30
Bullen, K. E., 86, 96–97, 98

Clairaut, A. O., 106
Coefficient of flattening, 115
Compressional waves, 68
Continental crust
    constitution of, 98
    density, 3
    mass, 3
    thickness (radius), 3
Continental drift, 37–39
Continental rocks, 9
Core, the, 2–3
    constitution of, 102–104
    density, 3
    depth to interface, 93

mass, 3
outer part of, 28–29
thickness (radius), 3
Cratons (shields), 9–10
Crust, the, 2–3
    constitution of, 98
    density, 3
    depth to interface, 93
    mass, 3
    thickness (radius), 3
Curie point, 30–31

Declination, 34, 35
Densities, 2–3
    distribution, 98, 99
    of minerals, 99,
    of simple oxides, 101
Depth of compensation, 109
Dipole, magnetic field of,
    26–27
Distortional waves, 68
Dynamo theory, 28–30

Earth Materials (Ernst), 2,
    65, 99–100, 102
Earthquakes, 1, 53–57, 67,
    73–77
    deep-focus, 76–77
    distribution of, 7, 8, 54–56
    in rift valleys, 24
Eicher, D. L., 2, 42
Elastic ratios of simple
    oxides, 101
Elevations, 3–5
Elsasser, W. M., 28–30
Epicenter, the, 67
Equipotential (level) sur-
    faces, 106
Ernst, W. G., 2, 65, 99–100,
    102
Erosion, 10
    mountains and, 1–2
Everest, Sir George, 109

Fault-block mountains,
    21–22
Faults, 10–11
    motion of, 10

normal, 10, 11
overthrust, 10, 11
reverse, 10, 11
thrust, 10
transcurrent, 10, 15–18, 48
transform, 48–49
Ferrimagnetic materials, 30,
    31
Ferromagnetic materials, 30,
    31
Focus (focal) region, 67
Folds, 10–11
    recumbent, 11, 12, 14
Free oscillations, 69–73
Free-air anomaly, 107
Free-air correction, 107

Geodesy, 105
Geoid, the, 106, 107
    undulations of, 115, 116
Geologic time, 1–2
Geologic Time (Eicher), 2,
    42
Geosyncline, the, 57–60
Geothermal gradient, 119,
    120
Goldschmidt, V. M., 102
Graben, 23
Gravity
    anomalies, 106–108
    defined, 106
    direction of, 106
    force of, 106
Gravity field, 105–117
    departures from isostatic
        compensation, 112–
        114
    shape of the earth, 114–
        117
    theory of isostasy,
        108–112
Gravity-sliding tectonics, 15
Gregory, J. W., 23
Gutenberg, Beno, 85–89, 92

Hall, James, 57
Hayford, J. F., 109

Heat flow, 118–126
  terrestrial, 119–121
Heiskanen, W. A., 110
High-order modes, 71–72
Hubbert, M. K., 15
Hydromagnetics, 29

Inclination, 34, 35
Inhomogeneity of the mantle, 95–98
Internal divisions, 2–3
Internal friction, 90
International Gravity Formula, 107
Isostasy, theory of, 108–112
Isostatic compensation, 109
  departures from, 112–114
Isothermal remanent magnetization, 31
Isotropic materials, 68

Jeffreys, Sir Harold, 86–89, 92, 95

Lehmann, I., 85, 86
Lines of force, 26
Lithosphere, 65
Longitudinal waves, 69
Love, A. E. H., 73
Love waves, 73

Magnetic field, 26–45
  and continental drift, 37–39
  of a dipole, 26–27
  dynamo theory, 28–30
  isothermal remanent, 31
  lines of force, 26
  paleomagnetism, 31–37
    defined, 31–32
    three assumptions of, 32–33
  reversals of, 30
  rock magnetism, 30–37
    declination, 34, 35
    inclination, 34, 35
    stability, 32
  secular variation, 27
  self-reversals of, 31
  thermoremanent, 31
  Vine-Matthews hypothesis, 39–45
  westward drift, 27–28
Magnetohydrodynamics, 29
Mantle, the, 2–3
  constitution of, 98–102
  density, 3
  depth to interface, 93
  inhomogeneity of, 95–98
  mass, 3
  thickness (radius), 3
Mass, 3
Matthews, D. H., 39–45
Meinesz, F. A. Vening, 114

Mobile belts, 9, 10
Modes
  high-order, 71–72
  normal, 69, 70
  radial, 70, 71
Moho, the (Mohorovičić discontinuity), 2, 50
Mountains, 11–23
  building, 57–60
  erosion and, 1–2
  fault-block, 21–22
  roots of, hypothesis, 109

Nappes, 11–15
  structure sections, 19–22
Newton, Sir Isaac, 106
Nodes, 70
Normal faults, 10, 11
Normal modes, 69, 70

Oceanic crust
  density, 3
  mass, 3
  thickness (radius), 3
Oceans, 2
  density, 3
  elevations in, 3–5
  mass, 3
  ridges, 49–53
  trenches, 53–57
Oceans (Turekian), 42, 46, 120
Orogenic belts, 9
Orogeny, 9
Oscillations
  free, 69–73
  spheroidal, 71
  torsional, 71, 72
Overthrust faults, 10, 11

Paleolatitudes, 35, 36–37
Paleomagnetism, 31–37
  defined, 31–32
  three assumptions of, 32–33
Paramagnetic materials, 30–31
Period of vibration, 69
Phase transformations, 98
Plate tectonics, 46–66
  driving mechanism, 64–66
  motions of present-day plates, 60,–64
  and mountain building, 57–60
  ocean ridges, 49–53
  transform faults, 48–49
Pratt, Archdeacon, 109
Pratt-Hayford scheme of compensation, 109–112
Precession, 117
Pressures in the Earth, 93–95

P-waves, 68–69, 73–76, 78
  push-pull motion of, 69
  travel times, 83–89

Radial modes, 70, 71
Radioactivity, heat generation by, 121–124
Rayleigh, Lord, 73
Rayleigh waves, 73
Rays, 78
Receding branches, 80
Recumbent folds, 11, 12, 14
Refraction, 78–79, 81
Regenerative dynamo, 29–30
Reverse faults, 10, 11
Ridges, 49–53
Rift valleys, 23–25
Ringwood, A. E., 65, 103
Rocks
  age of, 9, 23
  bedding, 10–11
  formed by metamorphism of basalt, 65–66
  magnetism, 30–37
    declination, 34, 35
    inclination, 34, 35
    stability, 32
  sedimentary, 10–11, 13, 22
Roots of mountains hypothesis, 109
Rubey, W. W., 15

Sea-floor spreading, 43–47
Secular variation, 27
Sedimentary rocks, 10–11, 13, 22
Seismic attenuation, 90–91
Seismic waves, 68–69
Seismology, 67–91
  behavior of body waves at an interface, 77–83
  defined, 67
  free oscillations, 69–73
  seismic attenuation, 90–91
  seismic waves, 68–69
  travel times, 83–90
    body waves, 73–77
  velocity distribution, 83–90
Shadow zones, 80
Shape of the Earth, 114–117
Shear waves, 68
Shock waves, 95
SH-waves, 69, 73
Snell's Law, 78
Solar System, The (Wood), 102
Spheroid, the, 106
Spheroidal oscillations, 71
Surface, the, 3–5
Surface of the Earth, The (Bloom), 1
Surface waves, 69
SV-waves, 69, 73

S-waves, 68–69, 73–76, 78
  shake motion of, 69
  travel times, 83–89

Tectonic processes, 9
  gravity-sliding, 15
  plate, 46–66
    driving mechanism,
      64–66
    motions of present-day,
      60–64
    mountain building,
      57–60
    ocean ridges, 49–53
    transform faults, 48–49
Temperatures, 118–126
  deep in the Earth,
    124–126
  heat generation by radio-
    activity, 121–124
Terrestrial heat flow,
    119–121
Thermal conductivity, 119
Thermoremanent magnetiza-
    tion, 31
Thrust faults, 10
Torsional oscillations, 71, 72
Transcurrent faults, 10,
    15–18, 48

Transform faults, 48–49
Transition zones, 92
  constitution of, 101–102
  depth to interface, 93
Transverse waves, 69
Trenches, 53–57
Turekian, K. K., 42, 46, 120

Valleys
  grabens, 23
  rift, 23–25
Vector fields, 26
Velocities
  distribution of, 83–90
  of minerals, 99
Vertical exaggeration, 11
Vibrations
  normal modes of, 69, 70
  period of, 69
Vine, F. J., 39–45
Vine-Matthews hypothesis,
    39–45
Volcanoes, 1, 20–21, 57
  distribution of, 5, 6
  in rift valleys, 24

Wave front, the, 68
Waves
  body, 69

behavior at an interface,
    77–83
  travel times, 73–77
  compressional, 68
  distortional, 68
  longitudinal, 69
  Love 73
  P-, 68–69, 73–76, 78
    push-pull motion of, 69
    travel times, 83–89
  primary disturbances, 68
  Rayleigh, 73
  S-, 68–69, 73–76, 78
    shake motion of, 69
    travel times, 83–89
  seismic, 68–69
  SH-, 69, 73
  shear, 68
  shock, 95
  surface, 69
  SV-, 69, 73
  transverse, 69
Wegener, Alfred, 38
Westward drift, 27–28
Williamson-Adams equation,
    95–97
Wilson, J. T., 48
Wood, J. A., 102

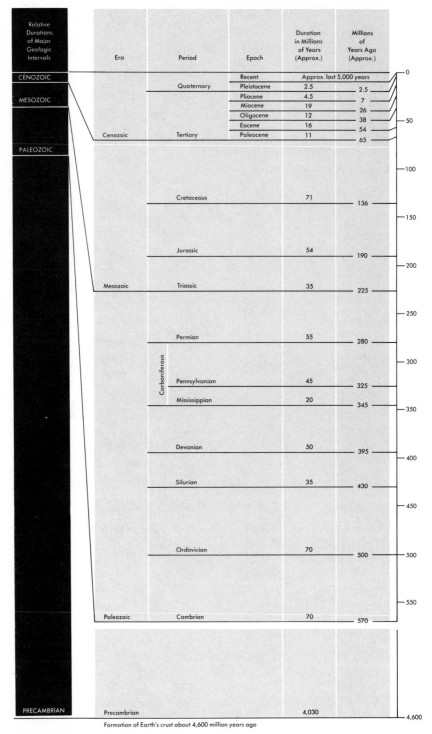

Relative Durations of Major Geologic Intervals	Era	Period	Epoch	Duration in Millions of Years (Approx.)	Millions of Years Ago (Approx.)
CENOZOIC		Quaternary	Recent	Approx. last 5,000 years	0
			Pleistocene	2.5	2.5
MESOZOIC			Pliocene	4.5	7
			Miocene	19	26
			Oligocene	12	38
			Eocene	16	54
	Cenozoic	Tertiary	Paleocene	11	65
PALEOZOIC		Cretaceous		71	136
		Jurassic		54	190
	Mesozoic	Triassic		35	225
		Permian		55	280
		Carboniferous Pennsylvanian		45	325
		Mississippian		20	345
		Devonian		50	395
		Silurian		35	430
		Ordovician		70	500
	Paleozoic	Cambrian		70	570
PRECAMBRIAN	Precambrian			4,030	4,600

Formation of Earth's crust about 4,600 million years ago

Millions of Years

Radiometric ages after Harland, Smith, and Wilcock (eds.), 1964, *The Planerozoic Time Scale*. part 2: *Radiometric methods with respect to the Time Scale*. Geol. Soc. London, v. 120 S.